ROUTLEDGE LIBRAR
POLITICAL GEO(

Volume 7

DEVELOPMENTS IN ELECTORAL GEOGRAPHY

DEVELOPMENTS IN ELECTORAL GEOGRAPHY

Edited by
R.J. JOHNSTON, F.M. SHELLEY AND
P.J. TAYLOR

Routledge
Taylor & Francis Group

LONDON AND NEW YORK

First published in 1990

This edition first published in 2015
by Routledge
2 Park Square, Milton Park, Abingdon, Oxon, OX14 4RN

and by Routledge
711 Third Avenue, New York, NY 10017

Routledge is an imprint of the Taylor & Francis Group, an informa business

British Library Cataloguing in Publication Data
A catalogue record for this book is available from the British Library

ISBN: 978-1-138-80830-0 (Set)
eISBN: 978-1-315-74725-5 (Set)
ISBN: 978-1-138-80991-8 (Volume 7)
eISBN: 978-1-315-74974-7 (Volume 7)
Pb ISBN: 978-1-138-80995-6 (Volume 7)

Publisher's Note
The publisher has gone to great lengths to ensure the quality of this reprint but
points out that some imperfections in the original copies may be apparent.

Disclaimer
The publisher has made every effort to trace copyright holders and would
welcome correspondence from those they have been unable to trace.

Printed and bound by CPI Group (UK) Ltd, Croydon, CR0 4YY

Developments in Electoral Geography

edited by
R.J. JOHNSTON, F.M. SHELLEY,
and
P.J. TAYLOR

Routledge
London and New York

First published 1990
by Routledge
11 New Fetter Lane, London EC4P 4EE

Simultaneously published in the USA and Canada
by Routledge
a division of Routledge, Chapman and Hall, Inc.
29 West 35th Street, New York, NY 10001

© 1990 R.J. Johnston, F.M. Shelley and P.J. Taylor

Typeset by LaserScript Limited, Mitcham, Surrey

Printed and bound in Great Britain by
Billings & Sons Limited, Worcester

British Library Cataloguing in Publication Data

Developments in electoral geography.
  1. Politics. Geographical factors
  I. Johnston, R. J. (Ronald John, 1941–) II. Shelley, F. M. 1947–
  III. Taylor, Peter J. (Peter James, 1944–)
  320.1'2

ISBN 0–415–04133–3

Library of Congress Cataloging in Publication Data

Developments in electoral geography / edited by R.J. Johnston, F.M. Shelley,
  and P.J. Taylor.
      p.   cm.
  Bibliography: p.
  Includes index.
  ISBN 0–415–04133–3
  1. Elections. 2. Geography, Political. I. Johnston, R. J. (Ronald John) II.
  Shelley, F. M., 1947– . III. Taylor, P. J., 1944– . IV Title: Electoral
  geography.
  JF 1001.D47 1990
324–dc20
89–10562
CIP

Contents

Tables

Figures

Contributors

R. J. Johnston Department of Geography, University of Sheffield, Sheffield S10 2TN, UK.

F. M. Shelley Department of Geography, University of Southern California, University Park, Los Angeles, CA 90089-0063, USA.

P. J. Taylor Department of Geography, University of Newcastle upon Tyne, Newcastle upon Tyne, NE1 7RU, UK.

J. A. Agnew Department of Geography, Syracuse University, Syracuse, NY 13244, USA

N. P. Passchier and H. H. van der Wusten Department of Geography, University of Amsterdam, Jodenbreestraat 23, 1011 NH Amsterdam, The Netherlands.

J. O'Loughlin Department of Geography, University of Colorado, Boulder, CO 80309-0260, USA.

A. J. Parker Department of Geography, University College Dublin, Belfield, Dublin 4, Irish Republic.

R. Honey Department of Geography, University of Iowa, Iowa City, IA 52242, USA.

J. R. Barnett Department of Geography, University of Canterbury, Private Bag, Christchurch, New Zealand.

M. Eagles Department of Political Science, Saint Mary's University, Halifax, Nova Scotia B3H 3C3, Canada.

A. Lijphart Department of Political Science, University of California – San Diego, La Jolla, CA 92093- 0060, USA.

R. S. Roberts and F. M. Ufkes Department of Geography, University of Iowa, Iowa City, IA 52242, USA

D. C. Hodge and L. A. Staeheli Department of Geography, University of Washington, Seattle, WA 98195, USA.

D. Whalley Department of Geography and Anthropology, University of North Texas, Denton, TX 76203–5277.

B. Grofman and A. Glazer School of Social Sciences, University of California – Irvine, Irvine, CA 92717, USA.

R. Griffin Department of Political Science, Stanford University, Stanford, CA 94305, USA.

N. Ettlinger Department of Geography, Ohio State University, Columbus, OH 43210, USA.

A. B. Murphy Department of Geography, University of Oregon, Eugene, OR 94703, USA.

G. L. Clark School of Urban and Public Affairs, Carnegie-Mellon University, Pittsburgh, PA 15213, USA.

Chapter one

Developments in electoral geography

Fred M. Shelley, R. J. Johnston
and Peter J. Taylor

Electoral geography, or the systematic spatial analysis of elections, has
a long intellectual history. The mapping of voting statistics has provided
important insights into the operation of modern polities. But such
contributions have been rare. Electoral geography has been a victim of
that most debilitating of all intellectual diseases – rampant empiricism.
Many electoral geography studies have consisted of mere descriptions
of the spatial pattern of the vote in a particular election with little or no
concern for wider issues. Hence the social theory underlying the
analysis is left implicit as a *status quo* model of society. This has
generated a very uncritical body of knowledge under the label electoral
geography.

It was a generally felt dissatisfaction with this intellectual legacy that
provided the stimulus for organizing a conference on electoral
geography and social theory in Los Angeles in April 1988. The purpose
was to slay the empiricist dragon once and for all and the results of this
exercise are before you. What is very clear is that electoral geography is
in a state of flux. In this it reflects the current pluralism of human
geography as a whole. There is a sense of transition from empiricism but
it is by no means obvious where we are going. A diversified product is
to be welcomed as long as we avoid an eclectic blurring of theory. At
Los Angeles the focus on social theory was manifest in two ways. First
some papers provided critical evaluations of traditional theories in
recent electoral contexts. Second, some papers attempted to look
outwards to alternative social theories in which to situate electoral
geography. Between them these approaches combine continuity and
innovation to provide electoral geography with a fresh platform for
researches in the 1990s.

Electoral geography's unsatisfactory legacy is largely due to a recent
history of intellectual neglect. In the early twentieth century, electoral
geography played a major role in the development of social science
theory. For example, the American historian Frederick Jackson Turner
relied heavily on mapped election results in developing his theses about

the roles of sectionalism and the frontier in American history (Turner, 1908).

The intellectual prominence of electoral geography was reduced, however, following the Second World War. The mid-twentieth century was characterized by a number of related developments associated with the achievement of American hegemony over the world economy, including the development of a liberal, pluralist conception of the democratic political order, plus methodological and technical advances in social science which encouraged researchers to rely on individual-level survey data in empirical research. Behaviouralism, which bases empirical research on analysis of the reported attitudes and characteristics of individual voters, came to dominate political science in the 1950s. Behavioural analysis relied on data derived from individual survey research, which was facilitated by sophisticated polling methods that provided inferences about an entire population from small samples of voters. Thus researchers examined relationships between reported voting behaviour and voter characteristics such as economic status, income, education, ethnicity, religion and occupation.

The behavioural approach is closely linked to a liberal, pluralist conception of democratic governance. Pluralism in democracy implies that governments are elected following open competition among organized political parties which offer alternative platforms to rational, self-interested voters. Implicit in the assumption of voter rationality is the voter's belief that exercise of the franchise can lead to increases in individual utility (Downs, 1957). Thus voters select the party whose proposals are most consistent with their own views. Individual voter utilities and hence actual votes can be predicted on the basis of income, ethnic status and other economic and social characteristics. This pluralist conception of democracy was assumed to characterize Europe, North America and other industrialized democracies. Moreover, in conjunction with contemporary development theory pluralism implied that an important component of the 'modernization' process was the eventual adoption of Western democratic values in developing nations.

Critiques of electoral geography

The dominance of the behaviouralist paradigm in political science relegated electoral geography to a secondary intellectual position during the 1950s and 1960s. Only as the ideology of liberal pluralism came to be questioned did electoral geography begin to return to its former intellectual stature. These historical considerations provide a backdrop for the first section of the book. The chapters by John Agnew and David Reynolds trace the intellectual development of electoral geography since the 1960s. Recognizing that the selection of governments hinges upon decisions made by individual voters and the procedures by which the voters' preferences are aggregated, both Agnew and Reynolds identify several distinct lines of enquiry focused on the geography of electoral behaviour and of electoral systems.

Agnew briefly reviews research that has been undertaken from what he terms 'modernization–nationalization' and 'social welfare' approaches. Both of these approaches represent the philosophical underpinnings of the paradigm of liberal pluralism. Indeed, the modernization-nationalization perspective is a statement of the relationships between liberal pluralism and contemporary regional development theory, as summarized in Table 2.3. More directly relevant from the perspective of social theory are two alternative approaches examined by Agnew: the perspective of uneven development inspired by the world-systems analyses of Wallerstein and that of place context, which examines electoral geography in terms of daily experience within social contexts associated with particular places. For Agnew, the latter two approaches are not only more consistent with the mainstream of current social theory, but they are also indeed more geographical and hence more consistent with the application of social theory in human geography.

Reynolds, like Agnew, identifies the uneven development and place-context approaches as major areas for intellectual development in electoral geography. He also points out the intellectual linkages and analogies between the world systems framework and the politics of place. While the former concentrates on global issues, the latter emphasizes local concerns. However, at both scales political competition is bound up in struggles between factions of capital as well as between capital and labour. The world-systems perspective identifies a single global economy composed of discrete nation-states which compete for influence within the world economy. The position of each nation-state within the world economy affects its internal politics and indeed politicians often debate policies intended to 'improve' their country's position. American politicians, for example, campaign to improve the prospects for competing with Japan, Russia, Western European, and other nations. At the local scale, analogous patterns of competition between factions of capital can often be identified and may be manifested in land use or other locational conflicts. Thus, as Reynolds indicates, place-related contextual information can help to account for the frequent observation that sectional and regional conflicts continue to be important to the outcomes of national elections in many countries.

The substantive contributions in Parts II and III of this book illustrate many of Agnew's and Reynolds's arguments. Part II comprises a selection of case studies of elections in countries where the analyses follow the modernization-nationalization tradition of the Lipset and Rokkan social cleavage model. Part III deals with elections in the United States and is mainly concerned with place-based politics and local issues. Finally, Part 4 contains three chapters that focus on new directions for study in electoral geography.

The cleavage model and its implicit social theory

The role of organized political parties and cleavages along party lines among national or subnational electorates is the underlying theme of the second part of the book, which includes chapters on electoral geography in other developed countries of the world. The theme of party and electoral cleavage is discussed in the chapter by Ron Johnston, who addresses the issue of the continuing role of electoral cleavages in post-industrial society. Traditional electoral theory as articulated by Lipset and Rokkan (1967) posits the existence of long-standing cleavages between groups of voters, with political parties mobilized on either side of the cleavage. According to this model, parties in Western democracies tend to be organized along one or more of four types of cleavage: between centre and periphery; between urban and rural interests; between church and state, or between different religious interests; and between labour and capital. Generally the development of centre-periphery and religious-based cleavages antedated the Industrial Revolution while the latter developed after transformation to an industrial economy. Johnston addresses the validity of this model by pointing out the increasing phenomenon of *dealignment*, in which traditional loyalties between party and voter are weakened. Dealignment appears to cast doubt on the longevity of the traditional cleavages posited by Lipset and Rokkan (Johnston, 1987). In examining recent electoral shifts in Great Britain, Johnston points out that while in general cleavages act to structure political conflict, cleavages that are too rigid frustrate the possibility of minority victory, creating long-run instability in the country's political system. Thus long-run stability requires short-run flexibility. As political parties can and do attempt to manipulate cleavages for electoral benefit, further changes in alignment patterns are likely, not only in the United Kingdom but elsewhere as well.

The 1987 general election in New Zealand illustrates the impact of realignment based on the restructuring of traditional political cleavages. This election is discussed in the chapter by Rex Honey and Ross Barnett. Over the past several decades, the urban-based Labour Party has competed with the more conservative and more rural-oriented National Party for influence in New Zealand. The concentration of Labour supporters in the nation's larger cities has aided National owing to an electoral bias resulting from the concentration of Labour supporters in urban districts (Johnston, 1976). In most elections, Labour must win a substantial majority of votes cast nation-wide in order to ensure a parliamentary majority. This situation changed, however, in 1987. By adopting conservative fiscal policies emphasizing the free market, the incumbent Labour government was able to reach out and increase its support by cutting across traditional cleavage lines. While the Labour government's new economic policies alienated some of the party's traditional supporters, such alienation was insufficient to overthrow the government. Encouraging such realignment in the electorate, Labour may have initiated a restructuring of New Zealand politics. Thus Honey and Barnett illustrate the importance of the role of the political party in realignment. In New Zealand, Labour's change in economic policy not only restructured the electorate, but also signalled a willingness to abandon the traditional class-based cleavage typical of developed British Commonwealth countries.

The importance of the traditional class cleavage in British politics is addressed by Munroe Eagles, who reviews the relative merits of individual-level as opposed to aggregate electoral research, pointing out that the two approaches tend to yield divergent conclusions. In Britain, for example, Eagles argues, survey research indicates a recent decline in class-based voting while aggregate data analysis documents continued class cleavages. Integrating these perspectives by reporting the results of an intensive study of working-class voters in Sheffield, he indicates support for the place-based approach to politics advocated by Agnew (1987) and Johnston (1985). Although persons in working-class neighbourhoods tended to vote in similar ways, it cannot necessarily be concluded that these similarities are the result of neighbourhood-level diffusion processes. Thus Eagles, like Johnston, concludes with advocacy of a view of local homogeneity that is 'set more firmly into the context of place-based socialization'.

Nico Passchier and Hermann van der Wusten illustrate the 'electoral dynamics and historical change' theme which Reynolds sees as becoming increasingly popular in electoral geography. Using standard factor analytic methodology, they define electoral eras and regions for the Netherlands over the period 1888–1986, and identify continuity in the traditional three 'pillars' of Dutch politics even after the secularization of social life, with the decline of the Christian Democrat pillar and the compensating renewal of the Liberal pillar.

Assessing the validity of the Lipset–Rokkan model for Ireland, John O'Loughlin and Tony Parker discuss the unique position of Ireland among its European neighbours. Within the European Community, Ireland is a peripheral nation-state, with traditional dependence on and animosity towards Great Britain in particular. Ireland was long a colony of Britain, and the aftermath of its struggle for political independence, including the island's division between the Irish Republic and Northern Ireland, which remained in the United Kingdom, has left a lasting impact on Irish politics. In particular, the cleavage between capital and labour characteristic of the politics of many European countries was sublimated to the national independence movement in the early twentieth century. Unlike Great Britain and New Zealand, members of the Irish Parliament are elected through an electoral system which results in substantially more representation of minority views and interests. These factors have influenced the outcomes as well as the distributions of votes in national referenda on a variety of economic and non-economic issues.

In commenting on the other chapters in this section, Arend Lijphart focuses on the Lipset–Rokkan cleavage model as articulated by Johnston and elaborated upon by the other authors. Lijphart notes that the four basic partisan cleavages postulated in the Lipset–Rokkan model are insufficient to deal with the complexities of contemporary Western democratic politics (as also do Harrop and Miller, 1987). Thus he proposes the addition of several additional dimensions – foreign policy cleavages, regime support (including the role of parties on the extreme right or left of a national political spectrum), growth of the governmental sector, and post-industrialism. The last three, which together take into account the issues of participatory democracy and environmentalism, do not fit neatly with the current left–right spectrum of internal politics in most Western nations. The Green Party of West Germany and similar parties elsewhere, as well as anti-governmental, libertarian-oriented political movements, can be accounted for in an expanded typology of this sort. This expansion can also help in responding to Johnston's call for increased flexibility and dynamism in political party organization. Indeed, such increased flexibility and dynamism may be crucial in the development of an increasingly international outlook associated with the increased economic and political co-ordination of contemporary western Europe.

The American experience and the ideology of democracy

The Lipset and Rokkan model has not been widely used in the analysis of United States elections; Taylor and Johnston's (1979) attempt to do this is untypical. Instead, the often-noted exceptionalism of US politics is reflected in geographical studies of elections there. The chapters in this part of the book reflect that, at both national and local scales. The United States is the oldest functioning liberal democracy in the world. In many respects American democracy is unique among Western liberal democracies. Compared to most European countries, American government is decentralized and territorially oriented, with strong linkages between representatives to national, state and local legislative bodies and their constituents. American electoral procedures emphasize place-related considerations, and sectional and territorial conflicts in American politics often eclipse those based on class cleavages, which tend to dominate the politics of European nations (Agnew, 1987; Shelley, 1988). Thus place context as identified by Agnew is particularly important in understanding the geography of American politics.

These unique characteristics of American democracy underlie the third part of the book. In particular, the chapters there emphasize conflict between places and economic sectors, illustrating the importance of the place-context perspective described by Agnew and Reynolds. Rebecca Roberts, Frances Ufkes and Fred Shelley focus on a recent referendum on the legality of corporate farm ownership in the agricultural state of Nebraska. Application of social theory to the topic of agriculture in developed societies illustrates the contradictory class position of the Mid Western farmer at a time when the size and economic power of the agricultural sector in the developed world are declining rapidly. American farmers are on a 'technological treadmill'. Advances in production technology resulted in lower commodity prices, forcing farmers to expand production through either additional land acquisition or increased technical innovation, and these pressures have the long-run effect of increasing the efficiency of food production, driving prices down and placing renewed emphasis on increased efficiency. Responding to the 'technological treadmill' process, American farmers have long turned to electoral politics in attempts to reduce the impacts of large corporations on farming. However, empirical analysis of the referendum on corporate farming in Nebraska, undertaken from a social-theoretic perspective demonstrates that while place context and the effects of capitalistic differentiation could effectively predict the outcome of the election, its passage did little to influence the underlying structural relationships in the region's agriculturally based economy.

The chapters by David Hodge and Lynn Staeheli and by Diane Whalley focus on urban conflicts. Following the lead of Katznelson (1981), Hodge and Staeheli focus on the divergence between the politics of production and the politics of consumption. They argue that changing relationships between the workplace and the home associated with contemporary transition to a post-industrial economy have resulted in an increasing separation of contemporary consumption issues from partisan politics. In examining a series of ballot issues in the city of Seattle, Washington, Hodge and Staeheli identify two distinct dimensions – one which reflects partisan divisions in the electorate and emphasizes work-related issues and a second which is oriented to

7

consumption-based politics and is relatively independent of partisan elections.

Diane Whalley's chapter focuses upon the distribution of federal funds for housing rehabilitation in the city of Minneapolis, Minnesota. In particular, she focuses on the relationship between municipal electoral structures and policy outcomes in the context of the Community Development Block Grant (CDBG) programme, which is federally funded but locally administered. The CDBG programme requires local municipalities to identify target areas in which funds are to be concentrated. Areas chosen to receive such funds benefit not only from an influx of federal revenues but also from other beneficial changes resulting from a restructuring of local land-use dynamics. While the programme was intended to concentrate funds in low and moderate-income urban neighbourhoods characterized by sub-standard housing, in practice the criteria used to identify the neighbourhoods are highly politicized and subject to influence by the territorially based and decentralized system used to elect city officials in Minneapolis and elsewhere.

In commenting upon these three chapters, Nancy Ettlinger focuses on an issue raised implicitly in each chapter: the meaning of voting in its local context. In particular, she emphasizes the degree to which voting can be perceived as a means of instituting meaningful social change. In doing so, she illustrates the common ground to be found among the somewhat diverse place contexts represented by the three studies. Ettlinger's comments emphasize the importance of the ideology of liberal democracy in the United States. Such ideology stresses that through electoral politics individual groups or blocs of voters can attain meaningful influence over the public policy process, thus effecting social change. Passage of an initiative, for example, would institutionalize a new policy throughout an entire region. However, voters in some localities within the state, county or city might oppose this new policy, so is the electoral process biased by the definition of formal regions? Would a different administrative map produce a different geography of electoral outcomes? These considerations affect the differentiation between the politics of production and the politics of consumption posited by Hodge and Staeheli.

In contrast to the other chapters in this part which focus on local political issues, that by Bernard Grofman, Robert Griffin and Amihai Glazer focuses on Congressional politics in the United States. Examining roll-call votes on major issues in the United States Senate, they examine the degree to which senators of different parties from the same state vote similarly. Unlike their European counterparts, American legislators are not subject to party discipline, nor are they expected to adhere to party positions on major issues. Thus measuring convergence in voting patterns among senators of opposite parties from the same state represents an interesting test of the extent to which place-based politics as opposed to party loyalty affects voting patterns. The economic theory of democracy espoused by Downs (1957) suggested the likelihood of convergence to a single position in any state regardless of party affiliation, but the empirical results presented in the chapter suggest otherwise: party loyalty transcends state consistency in most cases.

Whither electoral geography?

The final part of the book explicitly addresses the question of where electoral geography should be heading. The first two chapters draw on European and United States case studies respectively, thereby continuing the distinction introduced in the first two parts. The last chapter suggests a way forward that takes us beyond those two areas. All three are concerned with extending the coverage of electoral geography beyond its traditional intellectual domain.

Alex Murphy develops the arguments of Agnew and Reynolds that electoral geographers should be more concerned with the role of 'place' in the electoral process. He introduces the concept of an ideology of place that is constructed (and is continually being reconstructed) in the practice of politics. This incorporation of place into electoral geography is seen as a formidable task, but an illustration of how it can be done is provided by a study of changing Belgian electoral politics, where a place-based, 'linguistic' politics has both reinforced and yet destroyed Lipset and Rokkan-type cleavages. Murphy concludes that perceptions of place are integral to a truly geographical study of elections.

In the second chapter Gordon Clark draws attention to the issue of the integrity of the electoral process. He advocates emphasis on electoral regulation to parallel those on electoral behaviour and electoral structure. The regulatory process in effect defines the ground rules under which electoral competition is held. Within its ideology, regulation implies the notion of fairness, without which liberal democracy cannot be expected to work. Reviewing his extensive contributions to the literature on labour union representation elections, Clark points out that in recent years the influence of the American labour movement has been in decline, in part as a result of the lessened importance of heavy industry in an increasingly post-industrial economy and in part because of recent laws and court decisions which have gone against unions in their disputes with the interests of capital. Three organizations – the National Labor Relations Board, the United States Department of Labor, and the federal judiciary – have contributed to law regulating the conduct of union representation elections, often in conflicting fashion. The interaction of these organizations with capital and labour, along with changes in the political climate regarding labour, has influenced election outcomes in many respects. Thus the regulatory process can have substantial impact not only on the outcome of individual elections but perhaps more importantly on the long-run relationship between capital and labour in the United States and possibly elsewhere. Clark argues that this approach should be developed to cover the more traditional concerns of electoral geographers.

Peter Taylor's chapter is devoted to extending electoral geography to the Third World. Only a minority of the world's population live in countries characterized by the tenets of liberal democracy. Taylor identifies three basic characteristics of liberal democracy: universal suffrage, pluralistic elections and political freedom, including the freedom to vote for any party or candidate. Liberal democracies can also be regarded as social democracies: that is, states characterized by welfare rights, income and wealth redistribution and a consensus that such redistribution is an essential function of governance. Liberal social democracies are by and large confined to the economic core of the European-centred world economy, and the welfare-state consensus in many countries developed in response to the threat of communism during the Cold War period of the 1950s. Taylor rejects the contention of pluralist-oriented development theorists that underdeveloped countries are evolving into liberal social democracies themselves. Studies of electoral processes in the Third World confirm this conclusion (Osei-Kwame and Taylor, 1984). Thus Taylor concludes that the task ahead for political geographers is to analyse the causal mechanisms relevant to the success and failure of liberal social democracy over space and through time.

In summary

As would be anticipated, no consensus on the way forward for electoral geography emerged among the participants of the Los Angeles conference, though there was clear consensus that a way forward was needed. That situation is reflected in the contributions to this book. All the authors represented here are seeking to extend the field of electoral geography, not merely into new empirical fields (important though that task is) but rather into a deeper theoretical involvement which both enriches the empirical analyses and enables electoral geographers to make a wider contribution to the understanding of contemporary politics.

Electoral geography is a small but lively sub-discipline within geography. The chapters of this book illustrate its vibrance, bringing together most of the leading practitioners in a series of contributions that identify the research frontiers and indicate ways forward.

References

Agnew, J. A. (1987) *Place and Politics:the Geographical Mediation of State and Society*. Boston, Mass.: Allen & Unwin.

Downs, A.. (1957) *An Economic Theory of Democracy*. New York: Harper & Row.

Harrop, M. and Miller, W. L. (1987) *Elections and Voters*. London: Macmillan.

Johnston, R. J. (1976) 'Spatial structure, plurality systems and electoral bias', *Canadian Geographer*. 10, 310–328.

—— (1985) *The Geography of English Politics: the 1983 General Election*. London: Croom Helm.

—— (1987) 'Dealignment, volatility and electoral geography', *Studies in Comparative International Development*, 22, 3–25.

Katznelson, I. (1981) *City Trenches: Urban Politics and the Patterning of Class in the United States*. New York: Pantheon.

Lipset, S. M. and Rokkan, S. (1967) 'Cleavage structures, party systems and voter alignments', in S. M. Lipset and S. Rokkan (eds.) *Party Systems and Voter Alignments: Cross-national Perspectives*. New York: Free Press 3–64.

Osei-Kwame, P. and Taylor, P. J. (1984) 'A politics of failure: The political geography of Ghanaian elections, 1954–1979', *Annals of the Association of American Geographers*, 74, 574–89.

Shelley, F. M. (1988), 'Structure, stability and section in American politics', Political Geography Quarterly 7: 51–8.

Taylor, P. J. and Johnston, R. J. (1979) *Geography of Elections*. Harmondsworth: Penguin.

Turner, F. J. (1908) 'Is sectionalism in America dying away?', *American Journal of Sociology* 13, 687–706.

Part I

The state of electoral geography

Chapter two

From political methodology to geographical social theory? A critical review of electoral geography, 1960–87

John A. Agnew

Research in electoral geography over the past thirty years, especially in Britain and the United States, has had a variety of objectives and perspectives. However, a methodological obsession, largely imported from American political science, has restricted the development of a distinctive geographical contribution. This chapter provides a brief description of the major objectives and the series of perspectives that have characterized the field, paying special attention to their political and geographical assumptions. A final section describes the major continuity in the field, a commitment to 'political methodology', and suggests that the present challenge is to move beyond this to consider the potential of the social theory which regards the geographical as intrinsic rather than epiphenomenal to the explanation of political behaviour.

The objectives of research in electoral geography

Over the years 1960–87 four objectives have been most important. The first is the geography of electoral behaviour. This is the same objective that has been pursued most consistently by the French school of electoral sociology (e.g. Siegfried, 1913; Goguel, 1983) (Table 2.1). The second is the effect of the geography of interpersonal information flow on individual voting behaviour. These two objectives have been by far the most important. Two others have received less attention. They are the geography of electoral systems (electoral districting, 'gerrymandering') and the relationship between the geography of electoral performance and the geography of organization and mobilization exhibited by political parties.

Perspectives

The four objectives have been examined from one or other of four perspectives (Table 2.2). Much more has been written on the first two objectives and these are emphasized in this section.

Table 2.1: The objectives of research in electoral geography

1	Geography of electoral behaviour (e.g. Busteed 1975; Brusa, 1984, 1986; Agnew, 1985
2	Geography of interpersonal information flow and voting behaviour (e.g. Cox, 1969; O'Loughlin, 1981)
3	Geography of electoral systems (e.g. Taylor and Gudgin, 1976; Johnston, 1982; O'Loughlin, 1982)
4	Geography of political parties and electoral geography (e.g. Osei-Kwame and Taylor, 1984; Johnston *et al.*, 1987)

Table 2.2 Perspectives in electoral geography

	Origin period	Perspective
I	1965–72	'Modernization–nationalization' (e.g. Cox, 1969; Busteed, 1975)
II	1972–74	'Demographic–welfare' (e.g. Cox, 1973; Johnston, 1979; Brunn, 1974)
III	1974–79	Uneven development, (e.g. Archer and Taylor, 1981; Johnston, 1977)
IV	1979–present	'Microsociological–place' (e.g. Taylor, 1985; Johnston, *et al.* 1987; Agnew, 1985, 1987a).

The first perspective is that of 'modernization–nationalization'. The classic example of this is Cox (1969), who placed his emphasis upon 'neighbourhood effects' (the effects of distance on information flows) and 'partisan cues' (information about parties) in order to explain electoral behaviour. Over time, Cox argued, there is a spatial homogenization of these geographical effects throughout a nation. Like the practitioners of American political science in the 1960s, Cox thought that modernization (industrialization and urbanization) produced a nationalization of electoral behaviour.

In the 1970s this perspective was added to by two others which were responses to the social problems 'discovered' at that time and the growth of regional political movements in Europe and North America. The first can be called the 'demographic–welfare' approach. Again, Cox provides an important example. In his book *Conflict, Power, and Politics in the City* (1973) Cox explained urban political behaviour in terms of the spatial distribution of social groups and their access to public goods and services. Brunn (1974) and Johnston (1979) provided similar arguments at the national rather than the urban scale.

The third perspective could be called 'uneven development'. It is largely neo-Marxist in inspiration. In the work of Archer and Taylor

(1981) the ideas of Frank and Wallerstein provide the framework for an explanation of the geography of American presidential elections from Andrew Jackson to Ronald Reagan. At the scale of the United States they find in the late nineteenth century a Republican 'core' and a Democratic 'periphery'; after the Great Depression a reversal takes place with the 'core' of the north-east Mid west becoming Democratic and the 'periphery' of the south and West becoming Republican. The uneven economic development of the country in relation to long-cycles of the world economy and internal migration patterns is identified as the most important cause of change.

Finally, the fourth perspective can be called 'micro-sociological-place'. This perspective is the most recent addition – but in a sense it is also the oldest. Why is there an apparent contradiction? Even though the emphasis now is often on the 'microsociological', the term 'place' provides a historical connection with the French school of social geography and its emphasis on geographical setting or milieu. Agnew (1985), for example, attempts to show that the pattern of support for the Scottish National Party (SNP) in Scotland is related to everyday experiences in the social contexts provided by particular places. All places have their own history, 'external' links and peculiar traditions. Consequently, places are the social contexts in which political expression is determined.

The political-geographical assumptions of the perspectives

Each of the perspectives has different assumptions (Table 2.3). The most important differences are (1) the scale of analysis (I = the individual and small groups; II = large groups; III = core and periphery zones; IV = place), and (2) the major causes of political behaviour (I = information; II = the public economy; III = the private economy; IV = everyday life).

Table 2.3: Political–geographical assumptions of the perspectives

I	Spatial diffusion of political consensus (nationalization of political behaviour); the effects of contagion and information flows
II	The effects of population composition; spatial access to public goods and services
III	The spatial relations between 'core' and 'periphery'; uneven development and cultural division of labour
IV	Everyday life in the economic-cultural contexts of diverse places; the historical development of places and 'sense of place'; 'active' socialization in place; party organization and place

Each of the sets of assumptions derives from a particular 'variety' of social science. For I it is the behavioural tradition of American political science, associated especially with the Michigan School of Voting Studies. For II it is the public choice approach from micro-economics or a Weberian sociology of welfare. For III it is neo-Marxist economics or world-system theory. IV draws on that recent work in sociology and human geography which regards political activity as socially constructed in places as a result of the interaction between people's everyday routines, on the one hand, and social and economic influences from other, more distant 'seats of power', on the other hand (Agnew, 1987a). There is a variety of ways in which IV can be put into practice, ranging from the emphasis on the 'macro' in Smith (1988) to the more micro-sociological emphasis of Mercer and Agnew (1988).

From political methodology to geographical social theory?

One continuity running through perspectives I to III has been the assumption that geography is either, as in perspectives II and III, epiphenomenal (i.e. the outcome of 'deeper' national or global economic or political processes) or, as in perspective I, residual (important at some early 'stage' in historical development but otherwise increasingly unimportant). Geographers have made either one or both of these assumptions because the political methodology they have used, based largely around the application of the general linear model to data organized by census categories (regression analysis, factor analysis), has led them to add distance as a variable *or* to use aggregate census data in a way that denies the pervasive geographical *constitution* of social processes. Only with the fourth perspective do we find the emergence of a view in which the geographical is intrinsic rather than epiphenomenal to explanation.

The key claim on behalf of a geographical social theory is that national social categories are neither empirically appropriate nor theoretically coherent as causes of voting behaviour. National categories cannot cause individual voting decisions. Causality can only be discovered as specific social mechanisms that translate social structure into individual acts and vice versa (Agnew, 1987a). From the point of view of this social theory, therefore, electoral geography needs to shift its focus from the psychology of individual voters and their census attributes to the history of places and parties (Gaxie, 1985; Johnston, 1985).

A second continuity lies in the presumption that empirical findings from application of the dominant methodology somehow 'speak for themselves' and make sense *because* of their origins in that methodology rather than because they make sense theoretically. Works as disparate in other respects as Cox (1968), O'Loughlin (1981), and Archer (1988) all share this faith in the grace of methodology. Something akin to what Primo Levi (1986, 148-9) notes of the practice of industrial chemists afflicts the practice of electoral geographers: 'with the passing of the years, what had been a crude measuring operation had lost its significance and was transformed into a mysterious and magical practice'.

A perhaps extreme example of why a shift in focus is so important is provided by the British political scientist McAllister. In a paper entitled 'Class, region, denomination and Protestant politics in Ulster', McAllister (1983) regresses a variety of social structure, religion and region variables on various attitudinal responses from a political survey. His results are statistically insignificant. He also reports R-squares from 2 to 7 per cent. He concludes (p. 283) 'the results reported here have shown that political differences within the Protestant community are not rooted in class, religion or region'. But at the same time even he finds these results not only surprising but also unintelligible. He continues, 'Regionalism has been shown [previously] to be an important and pervasive influence in social and political life ... In religion, denominational differences within Protestantism have strong support' (p. 283). However, rather than following up this commonsense observation with a plea for an alternative approach he reverts to form. Questioning his methodology is not on the cards. After all, that is all he has. He certainly has no coherent social theory to explain divisions in the Protestant community. As I have put it elsewhere in commenting on another paper by McAllister, 'There is no role in McAllister's analysis for context – individual interactions. His theorization of voting behavior will not allow it. The message determines the medium; the medium delivers the message' (Agnew, 1987b, 40).

A third continuity involves an obsession with methodological rather than substantive theoretical discussion. Detailed attention to questions of empirical measurement is certainly important. However, since so much rests in the dominant methodology on R^2s and significance test results there is perhaps an exaggerated tendency to overplay methodological discussion at the expense of developing (as opposed to borrowing) theory. This is especially apparent in many 'exchanges' and 'comments', beginning with the comments on Cox's 1968 article (see Taylor, 1969; Kasperson, 1969) and extending through the recent exchange between Johnston (Johnston and Pattie, 1987) and McAllister (1987).

Conclusions

Electoral geography has been characterized by a range of objectives and perspectives since its revival in the English-speaking world in the

1960s. The challenge facing the field today is to become more geographical. By this I mean it is past time to move on from the borrowed assumptions of most of the perspectives to the development of geographically informed social theory appropriate for addressing the various objectives of the field. This shift would also bring electoral geography into contact with a trend characteristic of human geography in general.

Note

1   A previous version of sections I–III of this chapter was presented to a conference on Political and Electoral Geography at the University of Parma (Italy), 28 May 1985. It was published as 'La geografia elettorale nel mondo angloamericano negli anni '70 e '80', in 1986.

References

Agnew, J. A. (1985) 'Models of spatial variation in political expression: the case of the Scottish National Party', *International Political Science Review* 6, 161–80.

—— (1987a) *Place and Politics: the Geographical Mediation of State and Society*. London: Allen & Unwin.

—— (1987b) 'Place, anyone? A comment on the McAllister and Johnston papers', *Political Geography Quarterly* 6, 39–40.

Archer, J. C. (1988) 'Macrogeographical versus microgeographical cleavages in American presidential elections: 1940–1984', *Political Geography Quarterly* 7, 111–25.

—— and Taylor, P. J. (1981) *Section and Party: Political Geography of American Presidential Elections from Andrew Jackson to Ronald Reagan*. Chichester: Wiley.

Brunn, S. D. (1974) *Geography and Politics in America*. New York: Harper and Row.

Brusa, C. (1984) *Geografia elettorale nell 'Italia del dopoguerra: edizione aggiornata ai risultanti delle elezioni politiche 1983*. Milan: Unicopli.

—— (ed.) (1986) *Elezioni, territorio, societa*. Milan: Unicopli.

Busteed, M. A. (1975) *Geography and Voting Behaviour*. London: Oxford University Press

Cox, K. R. (1968) 'Suburbia and voting behavior in the London metropolitan area', *Annals of the Association of American Geographers* 58, 111–27.

—— (1969) 'The voting decision in spatial context', *Progress in Geography* 1, 96–100.

—— (1973) *Conflict, Power, and Politics in the City: Geographic View*. New York: McGraw Hill.

—— (1979) *Location and Public Problems*. Chicago: Maaroufa.

Gaxie, D. (ed.) (1985) *Explication du vote: un bilan des études electorales en France*. Paris: Presses de la Fondation Nationale des Sciences Politiques.

Goguel, F. (1983) *Chroniques electorales: la cinquième republique après de*

Gaulle. Paris: Presses de la Fondation Nationale des Sciences Politiques.

Johnston, R. J. (1977) 'The geography of federal allocations in the U.S.', *Geoforum* 8, 319–25.

—— (1979) *Political, Electoral, and Spatial Systems*. Oxford: Clarendon Press.

—— (1982) 'Redistributing by independent commissions: a perspective from Britain', *Annals of the Association of American Geographers* 72, 457–70.

—— (1985) 'Class and the geography of voting in England: towards measurement and understanding', *Transactions of the Institute of British Geographers* 10, 2: 245–55.

—— and Pattie, C. J. (1987) 'Family background, ascribed characteristics, political attitudes and regional variations in voting within England, 1983: a further contribution', *Political Geography Quarterly* 6: 347–50.

—— O'Neill, A. B., and Taylor, P. J. (1987) 'The geography of party support: comparative studies in electoral stability', in M.J. Holler (ed.) *The Logic of Multiparty Systems*. Dordrecht: Martinus Nijhoff .

Kasperson, R. (1969) 'On suburbia and voting behavior', *Annals of the Association of American Geographers* 59: 405–11.

Levi, P. (1986) *The Periodic Table*. London: Abacus.

McAllister, I. (1983) 'Class, region, denomination and Protestant politics in Ulster', *Political Studies* 31, 275–83.

—— (1987) 'Comment on Johnston and Pattie', *Political Geography Quarterly* 6 351–54.

Mercer, J. and Agnew, J. A. (1988) 'Small worlds and local heroes: the 1987 general election in Scotland', *Scottish Geographical Magazine* 104: 136–45.

O'Loughlin, J. (1981) 'The neighbourhood effect in urban voting surfaces: a cross-national analysis', in A. D. Burnett and P. J. Taylor (eds.) *Political Studies from Spatial Perspectives: Anglo-American Essays on Political Geography*, Chichester: John Wiley.

—— (1982) 'The identification and evaluation of racial gerrymandering', *Annals of the Association of American Geographers* 72: 165–84.

Osei-Kwame, P. and Taylor, P. J. (1984) 'A politics of failure: the political geography of Ghanaian elections, 1954-1979', *Annals of the Association of American Geographers* 74: 574-89.

Siegfried, A. (1913) *Tableau politique de la France de l'ouest sous la troisième republique*. Paris: Colin.

Smith, N. (1988) 'The region is dead! Long live the region!', *Political Geography Quarterly* 7: 141–52.

Taylor, P.J. (1969) 'Causal models in geographic research', *Annals of the Association of American Geographers*, 59: 402–04.

—— (1985) *Political Geography: World-Economy, Nation-State and Locality*. London: Longman.

—— and Gudgin, G. (1976) 'The myth of non-partisan cartography', *Urban Studies* 13: 13–25.

Chapter three

Whither electoral geography? A critique[1]

David R. Reynolds

There is no doubt that electoral geography was a major, if not the major, growth area in political geography during the 1970s. As Taylor (1978, 153) has put it, 'elections are a positivist's dream'. They are productive of huge amounts of quantitative data in a form readily analysed via cartographic and statistical methods; data which ostensibly also possess considerable face-validity as reflections of the popular will on a key social issue – the selection of governments. Research in electoral geography has tended to focus on one of three issues: the *geography of voting*, where the objective is to explain the spatial pattern of voting in terms of some other mappable characteristic(s); *geographical influences on voting*, where the object is to explain voting (typically the decision-making of individual voters) on the basis of 'spatial' contexts; and the *geography of representation*, which explores the means through which votes are converted into 'seats' in alternative electoral systems. Excellent reviews of this work include Taylor and Johnston (1979), Taylor (1984b), and, on the geography of specifically American elections, Archer and Shelley (1985). To the three 'traditional' questions, Archer and Shelley make it clear that a fourth has been added to the research agenda of the field: a focus on electoral dynamics and historical change in the geographies of elections.

Although electoral geography has long been recognized as a sub-field of political geography, it has had little in common either theoretically or methodologically with the rest of the field. Typically the geography of elections is covered in one or two brief chapters near the end of textbooks in political geography, almost as an afterthought. As another case in point, consider the fact that Ron Johnston, certainly one of the major contributors to electoral geography over the past two decades, omits any mention of electoral geography in his essay on the political geographical nature of the state (Johnston, 1982). If political geography lacked a general theoretical coherence and any sort of explicit linkage to emerging social theory in the late 1970s, electoral geography certainly did too. Therefore the simple fact that several electoral geographers in the late 1970s began to explore ways of integrating research in electoral geography with the 'new' political geography must in itself be viewed as progress.

Much of the impetus underlying these explorations can be attributed to the emergence of a more explicit concern within electoral geography for understanding electoral change within the context of broader changes in the structure of societies. Attempts at integration have proceeded along two main lines: (1) efforts to situate political parties and elections within theories of the state and (2) efforts which have led to the rediscovery of the politics of locality or place as a means of explaining the persistence of regional and local variations in voting in Western democracies. These will be the focus of this critique.

Political parties, the state, and the two geographies of elections

Johnston (1979a, 1980a) was the first to attempt an integration of electoral and political geography by developing a variant of the input/output 'systems' model of Easton (1965). However, it was criticized by Taylor (1985, 153) for being uncritical and strongly ideological:

the systems approach has proved to be more important for what it reveals than what it solves. ... The assumptions upon which electoral geography has been built are laid bare. They turn out to be the classic liberal assumptions of the twentieth century core states: a receptive government responds to an electorate that articulates its demands through its representatives. All of a sudden, conflicts have disappeared, history is forgotten and political parties are nothing more than vehicles for transmitting candidate and voter preferences.

For Taylor, a more satisfactory integration required a much more critical appraisal of the roles of political parties in different societies and an explanation of why the holding of regularly scheduled elections occurs only in a minority of countries in the world (Taylor, 1984a; Osei-Kwame and Taylor, 1984; see also Johnston, 1984b). In his view, the world-systems approach of Wallerstein (1979) offered an appropriate framework for initiating such an integration (Archer and Taylor, 1981; Taylor, 1981, 1982, 1984a).

One attraction of the world-systems approach is that it is defined by two complementary structures each of which is necessary for the operation of the world economy: a capitalist mode of production for organizing economic competition and an inter-state system of sovereign states to organize the distortion or even mystification of that competition. Each state exists to relate its (territorial) fragment of capital to the larger world economy in such a way as to benefit 'national' capital differentially, but the success of a state in this endeavour is constrained by its structural position (core, periphery or semi-periphery) in the world-economy (Osei-Kwame and Taylor, 1984). In this view, the state is a capitalist state in the dual sense that it is part of a larger capitalist system and pursues policies designed to enhance capital accumulation by those capitalists who control its governmental apparatus. There are some strong parallels between this perspective on 'international' politics and Harvey's recent work on 'urban' politics (1985a) and the geopolitics of capitalism (1985b). A basic difference is that in Harvey's theory the ruling coalitions which form to maintain and enhance fixed investments of capital in particular places are cross-class alliances made up of both factions of capital and segments of labour, whereas in Taylor's world-systems perspective the dominant coalitions controlling states consist of particular factions of 'national' capital. Harvey's theory is specific to advanced capitalist states (those in the 'core' of the world-economy), whereas the world-systems framework is meant to elucidate politics in any part of the now global world economy.

Drawing on Gamble's (1974) earlier work on British politics, Taylor (1984a, 1985) introduces the concepts of a *politics of power* and a *politics of support*. Although he is vague on this point, it appears that the politics of power refers to competition between factions of capital, while the politics of support is more fundamentally concerned with class conflict and conflicts between segments of labour. In countries in the core, Taylor suggests political parties operate through these two sets of political processes, promoting 'national' capital accumulation in general through the politics of power and mobilizing mass support through the politics of support. He also suggests that there are two geographies underlying elections: a geography of power and a geography of support.

The geography of support is simply the geographical distribution of 'normal' party voting. It is more difficult to determine with any precision what Taylor means by the geography of power, but presumably it is a reflection of how a government must disburse public expenditure geographically in order to reward its major backers from among 'national' capitals. Essentially Taylor appears to accept a highly abstract and simplified theory of the state wherein the two essential functions of the state are to facilitate capital accumulation and maintain the legitimacy of capitalist social relations. The politics of power is the politics associated with accumulation and the politics of support is the politics of legitimation. According to Taylor a major task of a reconstituted electoral geography would be to relate these two geographies in particular national contexts.

Taylor (1984a) suggests that there is no general relationship between the two politics (and hence no invariant spatial relationship between the two geographies). However, he does list five propositions which he suggests can form the basis of a 'new' electoral geography:

1 There is no structural need for the two processes to provide a consistent basis for a party's actions. The processes occur in two distinct political arenas, one elite-based and the other popular-based.

2 There is a structural need to mystify the dual role of parties. Without this mystification the legitimation function of the party becomes impossible....

3 The balance between action based on one or other of the two processes will vary in terms of whether the party is in office or opposition. ...

4 The prime basis of action in the politics of power is class whereas there will be a variety of bases for the politics of support. ...

5 Parties may operate involutarily [sic] to deflect attention on contradictions away from the state by taking the blame for the situation: A potential crisis of the state becomes a crisis of the party. ... (p. 121).

One problem with these propositions is that their derivation is obscure; the last three appear to be historical/empirical observations, while the first and second are apparently meant to be theoretical statements derived directly from the necessary relations (or structural logic) of capitalism itself. However, since Taylor makes no effort to provide a derivation of them, either historically or structurally, they must be viewed as problematic, at least temporarily. For the first proposition to be consistent with the historical materialist perspective he is trying to develop, it should probably read: 'There is a structural need for the two processes *not* to provide a single consistent basis for a party's actions.' With this change, the second proposition simply becomes a corollary of it. Taylor is successful in identifying a number of historical examples to illustrate the descriptive veracity of his schema, but it needs much more theoretical elaboration in specific historical contexts if it is to provide a satisfactory linkage between electoral geography and political geography. Indeed, in their present state of development it is easy to criticize these ideas for being excessively 'structural', for falling prey to the 'errors of functionalism' and even for being ahistorical. It can also be argued that the founder of electoral geography, André Seigfried, had made similar observations about parties and elections in France as early as 1913 when he suggested that voters had little choice but to affirm general principles rather than pursue material interests in general elections (Sanguin, 1985).

Where Taylor's analysis does become more insightful theoretically is when he attempts to apply these ideas to elections in the Third World. In general, he argues that the structural position of most Third World countries in the world economy is such that no politics of power pursued by a party can lead to sustained capital accumulation for its supporters; hence the politics of support cannot be successful in legitimating the economic policies of a government or in legitimating the government itself. Instead, if elections are to continue to be the means of selecting governments, parties must continuously mobilize different sections of the electorate (geographies of support) (for details see Osei-Kwame and Taylor's 1984, case study of Ghana).

Although an important beginning, Taylor's attempt to integrate electoral geography and political geography through the introduction of world-systems concepts must be adjudged only a partial success. Part of the reason lies in his implicit treatment of parties as theoretical entities rather than as historical creations. Political parties are viewed as some sort of complex switching mechanism engaged in what might be called 'double duping'. On the one hand, parties are the dupes of various fragments of domestic capital – the politics of power – and on the other they are seen as cynically duping the masses in order to obtain and remain in power – the politics of support. Parties may be masters at organizing some issues into politics and others out, to paraphrase Schattschneider (1960), but they are unlikely to do so solely because of 'structural imperatives' (also see Cutter *et al.*, 1987). The more likely reason is to win elections without alienating their traditional supporters. The difficulty with Taylor's 'two politics' schema is that it fails to situate political parties in any coherent theoretical and/or historical context. As a theory of elections it is an improvement over that underlying liberal democracy, but as a theory of parties *and* elections it is far too mechanistic and abstract.

Taylor's schema can also be criticized for placing too much emphasis on the role of parties as mobilizers of non-class cleavages and not enough on class conflict in its 'politics of support' and for putting too much emphasis on conflict internal to the capitalist class in its 'politics of power'. In his work published prior to 1986 Taylor appears to have viewed successful parties as 'cadre parties' (those that 'evolved out of the factions and parliamentary groupings which were competing for control of the state apparatus before the "age of parties" and mass suffrage': 1986, S11). In his recent more historically oriented work drawing on the history of party systems, however, he has found it necessary to introduce 'mass parties' with their origins in class conflict into his conceptual schema as well (Taylor, 1986). Since historically these parties have attempted to mobilize working-class constituencies not already affiliated with cadre parties, they had no incentive to organize class conflict out of electoral politics (see, e.g. Johnston, 1979b, 1984b). The result (in terms of Taylor's schema) is that for mass parties it is quite possible for the politics of support and the politics of power to be congruent without creating crises in legitimation. The reason for this (at least theoretically) is quite simple: the politics of power in this case entails the formation of the equivalent of a national coalition consisting of factions and segments of both capital and labour. Although this may occur only in advanced capitalist countries, where class-struggle is sufficiently far advanced for the material interests of significant sections of labour to be congruent with those of factions of capital, it is a contingency that is not easily accommodated within Taylor's original schema.

Finally, a serious deficiency in Taylor's analysis, even with the introduction of a more historically sensitive treatment of parties, is that by retaining the concept of 'politics of support' it continues to bestow ontological privilege to particular 'cleavages' in an essentially *national* electorate rather than to particular places in a highly *regionalized* electorate. In other words, Taylor's analysis assumes that national parties, in their attempt to win elections, identify issues which appeal to particular segments of the electorate irrespective of where such segments may be located. This assumption is a questionable one empirically in respect of all the Western democracies, but particularly so in the US where recent research clearly indicates the persistence of sectionalism in voting patterns (see, e.g., Shelley, 1988). There is an alternative view of electoral politics (and by implication of electoral geography); one in which places provide an essential context in any explanation of social relationships including political behaviour. And a few geographers have made some considerable progress in accumulating evidence in support of it. Must place be taken seriously in a reconstituted electoral geography? This is the central question examined in the remainder of the chapter.

The politics of place

Throughout the 1960s and 1970s, the standard approach in explaining particular geographies of voting was to relate measures of the social composition in constituencies to some measure of partisan support within them. This is the so-called social cleavage model. Departures from 'model' expectations were frequently attributed to unspecified regional 'effects' or to underspecified sectional parochialisms. In states like the US where it was well known that the patterning of the relationships varied over large regions, models had to be calibrated on a regional basis. The prevailing view was that such regional effects, whatever their cause, were relatively minor and soon to be relegated to the 'scrap heap of history' as the voting response of the electorate became 'nationalized'.[2]

In general, electoral geographers accepted the dominant position of modern political sociology 'that political alignments have crystallized around *national* social cleavages to produce *national* patterns of political mobilization and partisanship' (Agnew, 1987a, 80). In those particular national contexts in which voting alignments varied regionally, the term 'sectional alignment' or 'sectionalism', was employed to describe the situation. Only to the extent that sectionalism persisted was it thought that place mattered (and even here place referred to rather large geographical regions).

As attention increasingly turned to changes in voting patterns, it became obvious that nationalization was not occurring in either of the two most studied national contexts – the United Kingdom and the United States – or in any of the other industrial democracies. Sectional alignments, where they existed, were changing, not disappearing (see, e.g. Agnew, 1988; Archer and Taylor, 1981; Archer and Shelley, 1985; Bensel, 1984) and smaller scale variations in electoral alignments were not only not disappearing but, in many cases, were actually increasing (see, e.g. Agnew, 1987a; Johnston, 1985, 1987a; Murauskas *et al.*, 1988). Elections were becoming volatile affairs often reflecting decreases in the strength of individual identifications with traditional parties but, with the exception of the US, this volatility did not reflect significant (or rapid) change in the spatial structure of existing alignments (Johnston, 1987b). These findings forced a re-examination of the nationalization thesis and the traditional approach to explaining the geography of elections. If sectionalisms and local variations in alignments were not disappearing, then they could no longer be ignored as minor aberrations. However, as both Agnew (1987a) and Johnston (1986d) have made clear, these findings, while disturbing to the old orthodoxy have not yet led to its abandonment, particularly in political science and political sociology.

The social cleavage model, wherein the geography of elections is accounted for by the geography of social class, continues to be 'protected', usually by appending 'fixing accounts' which supply *ad hoc* explanations of any 'residual' local effects. Electoral geography remains loath to abandon its heavy reliance on what Thrift (1983) has called a 'compositional' approach to the explanation of human behaviour in favour of a more 'contextual' approach. Johnston (1987b) has suggested that in a compositional approach a person's partisan attitudes are viewed as a function of her position within the division of labour, with the result that 'knowledge of a society's cleavage structure should be sufficient to provide excellent predictions of voting behaviour' (p. 10). Contextual explanation, on the other hand, argues in various ways for spatial location or place as the dominant influence on voting:

> Neither of these polar positions is empirically tenable, for every social location must occupy a physical (time-space) context – a locale in Giddens' (1984) terminology – and every context will contain various social locations. My claim here, however, is that to date too much attention has been placed on compositional approaches and too little on contextual.
>
> This claim is not that one's social location is unimportant as an influence on voting behaviour. It does argue, however, that much of one's understanding of that social location is learned in the local context.
>
> (Johnston, 1987b, 11).

The various 'fixing accounts' themselves fall into either compositional or contextual categories. In the compositional category are attempts to account for over and under predictions from the social cleavage model by selective migration (Curtice and Steed, 1982), or on the basis of consumption cleavages (Dunleavy, 1979, 1980). Also in this category are attempts to account for local variations from the social cleavage model on the basis of various forms of local party activities,[3] appeals to uneven spatial development and differentiation of areas into core and periphery (McAllister and Rose, 1984; Agnew, 1987a), and appeals to differences in local 'political culture' or local adaptations to a dominant ideology (Butler and Stokes, 1969; McAllister and Rose, 1984; Archer and Shelley, 1985; Johnston, 1986a; Agnew, 1987a; Savage, 1987). Each of these has met with some success in particular cases, but for a compositional approach to be successful, by definition it must be appropriate in all contexts.

There are far fewer types of 'fixing accounts' in the contextual category, but the few that exist have attracted much attention. By far the most frequently employed conceptual device is the so-called 'neighbourhood effect' of individual voting. This notion is usually attributed to the late V. O. Key (1949) (but is probably even older) and continues to be employed widely in electoral studies throughout all social science. Essentially the term is descriptive of the tendency for a dominant local opinion in a constituency either to enhance or to counteract the supposed 'effect' of social class (see, e.g. Ennis, 1962; Cox, 1968; Butler and Stokes, 1969). As such, the neighbourhood effect is dependent on a form of socio-spatial 'contagion' within constituencies (places) and seems to imply that increasing exposure to locally dominant ideas breeds a consensus around those ideas. This interpretation has been assailed on a number of grounds – the most damning of which is the inability of 'contagion' to be interpreted as the causal mechanism underlying the phenomenon (Dunleavy, 1979). Spatial propinquity is undoubtedly necessary for the effect to be perpetuated but that still begs the question of how locally dominant political attitudes are produced in the first place. In this sense, explanation by appealing to the neighbourhood effect only appears 'contextual', but is little different from any of the more standard 'fixing accounts' of the compositional variety. For the neighbourhood effect to be a plausible addendum to an otherwise macro-sociological explanation based on social-class cleavages, it must be linked to the historical development of politics in actual places. But, as Agnew (1987b, 40) has pointed out so persuasively:

> [If] one theorizes the causes of voting behavior microsocio-logically, in terms of the histories and social structures of specific places, then cross-sectional multivariate analysis misses precisely what should be emphasized: causality cannot be discovered as regularity independent of time and space but only as specific social *mechanisms* that translate social structure into individual acts and vice versa.

Agnew (1987a, b) argues for a fully contextualized view of political behaviour in which people who live in different places are led to vote in specific ways by being actively socialized into the different political attitudes depending on the social structure and historical antagonisms of those places. Similar views have also been expressed by Johnston (1986d, 1987b) and, to a lesser extent, by Taylor (1985). Appealing as these may be on ontological and methodological grounds, they are very recent and little research adopting the perspective has been accomplished to date. For the most part, these writers have tried to champion the need for the adoption of a place-based perspective on politics by reworking and reinterpreting the research findings of those employing more traditional approaches. This is particularly characteristic of the recent work of Agnew (1987a) and Johnston (1986a, b, c, d, e; 1987a, b). There are, however, two recent attempts to specify more precisely how a more fully contextualized analysis might proceed.

The first is Savage's (1987) attempt to theorize local political processes as a means of understanding political alignments in contemporary Great Britain. He reviews recent work on the geography of partisan support and concludes that increases in local variations in voting alignments have been caused neither by changes in the geography of social class nor by an intensification of local political culture. Instead, they are the result of similar 'types' of localities becoming increasingly *similar* in their voting patterns. While not rejecting the causal efficacy of social structure in accounting for voting, he argues that local social structures are not simply local fragments of an otherwise national social structure. The material interests of persons occupying different positions in the social structure are locally not nationally based. In particular, Savage suggests that how persons in different social positions vote is contingent on the performance or 'trajectory' of local labour and housing markets *vis-à-vis* other such markets, specifically in terms of employment and wage levels and the local market values of housing.

Although Savage's analysis is clearly specific to Great Britain and is highly tentative, he is able to present empirical evidence in support of his argument. The particulars of his argument are less important from our perspective than the basic thrust of his analysis. Rather than turning immediately to an historical–geographic analysis to develop some concept of local political ideology or political culture, he turns to a simpler, place-specific mode of analysis in which voting behaviour is caused, not directly by British or capitalist social structure, but by these social structures as they are mediated by materially significant aspects of place. Although Savage makes no attempt to link his analysis to Harvey's (1985a) theory of politics, the parallels between the two analyses are striking.

Next, mention must be made of Cox's (1987) critique of a paper by Johnston (1987b). Drawing upon Sayer (1984), Cox takes Johnston to task for his notion of 'compositional' forms of explanation in social science and its implicit conception of social structure as consisting only of relations between people based on the formal similarity of their enumerable characteristics rather than on the internal (or necessary) structure of relations between them.[4] Unlike Johnston, Cox subscribes to a particular place-based social theory in accounting for the geography of voting. He sees Harvey's theory of regional–class alliance formation (1985b) as providing an explanation of instabilities and dealignment in the geography of voting in advanced capitalist societies. The problem with Cox's view is that it seems to limit the theorization of the necessary relations underlying social life in real places under capitalism somewhat prematurely to only the capital–wage labour relation and the space–time contradictions of capitalist accumulation. Harvey's theory is probably better interpreted as providing a theoretical understanding for the necessity of a local politics based solely on the operation of spatially constrained class relations. But, as Harvey himself would agree, these class relations are insufficiently linked historically with other social relations (including those of gender, ethnicity, and racism) to produce a very satisfactory explanation of the concrete political reality characterizing real places. Harvey's theory is an important beginning, but it is still only a beginning.

Summary

A compelling case can be made for the necessity of incorporating the concept of place in explaining the geography of elections. The problem is in specifying more precisely how this might be accomplished without engaging in ontological pragmatics and/or in theoretical eclecticism. While it is certainly true that place or 'locale is where people learn their politics – at home and beyond' (Johnston, 1986d: 594), it is clear that a place-based contextual approach to understanding politics would bear little or no relationship to electoral geography as it is presently practised. Analysis would necessarily shift from the relatively narrow present emphasis on changes in voting alignments over time to deeper historical analyses of the exercise of power and social struggles against it in particular places during periods of significant social change. This is the direction research should take; traditional electoral geography must yield to a more thoroughly *political* geography.

Notes

1   An earlier version of this chapter was originally written as part of a larger piece analysing recent theoretical developments in Anglo-American political geography, particularly efforts to come to grips with recent advances in social theory (see Reynolds and Knight, 1989). For a detailed discussion of the 'new' political geography, this larger work should be consulted.

2   For an excellent discussion and analysis of the nationalization thesis, see Agnew (1987a).

3   The various party activities analysed include: agenda setting and local organization (Johnston, 1986d,e), local campaign spending (Johnston, 1986c), local variations in links to the workplace and/or control of local government (Johnston, 1986b, d), and the activities of 'third' parties (Curtice and Steed, 1982; Agnew, 1987a).

4   In some respects, Cox's concern here matches Savage's for identifying causal mechanisms. They differ, however, in that Cox appears concerned only with highly abstract causal mechanisms which are necessary, while Savage is more concerned with causal mechanisms which are concrete and either necessary or contingent.

References

Agnew, J. A. (1984) 'Place and political behaviour: the geography of Scottish Nationalism', *Political Geography Quarterly* 3: 151–65.

—— (1987a) *Place and Politics: The Geographical Mediation of State and Society*. London: Unwin Hyman.

—— (1987b) 'Place anyone?: A comment on the McAllister and Johnston papers', *Political Geography Quarterly* 6: 39–40.

—— (1988) 'Beyond core and periphery: the myth of regional

political-economic restructuring and sectionalism in contemporary American politics', *Political Geography Quarterly* 7: 127–39.

Archer, J. C. and Shelley, F. M. (1985) *American Electoral Mosaics.* Washington, D.C.: Association of American Geographers.

—— and Taylor, P. J. (1981) *Section and Party.* Chichester: Wiley.

Bensel, R. F. (1984) *Sectionalism and American Political Development 1880–1980.* Madison: University of Wisconsin Press.

Butler, D. and Stokes, D. (1969) *Political Change in Britain: Forces shaping Electoral Choice.* London: Macmillan.

Cox, K. R. (1968) 'Suburbia and voting behavior in the London metropolitan area', *Annals of the Association of American Geographers* 58: 111–27.

—— (1987) 'Comments on "Dealignment, volatility, and electoral geography"', *Studies in Comparative International Development* 22: 26–34.

Curtice, J. and Steed, M. (1982) 'Electoral choice and the production of government', *British Journal of Political Science* 12: 249–98.

Cutter, S. L., Holcomb, H. B., Shatin, D., Shelley, F. M., and Murauskas, G. T. (1987) 'From grass roots to partisan politics: nuclear freeze referenda in New Jersey and South Dakota', *Political Geography Quarterly* 6: 287–300.

Dunleavy, P. (1979) 'The urban basis of political alignment', *British Journal of Political Science* 9: 409–43.

—— (1980) 'The political implications of sectoral cleavages and the growth of state employment: Part 1, The analysis of production cleavages. Part 2, Cleavage structures and political alignment', *Political Studies* 28: 364–83, 527–49.

Easton, D. (1965) *A Systems Analysis of Political Life.* New York: Wiley.

Ennis, P. (1962) 'The contextual dimension in voting', in W. McPhee and W. Glaser (eds.) *Public Opinion and Congressional Elections.* New York: Free Press.

Gamble, A. (1974) *The Conservative Nation.* London: Routledge & Kegan Paul.

Giddens, A. (1984) *The Constitution of Society.* Berkeley and Los Angeles, CA.: University of California Press.

Harvey, D. (1985a) *The Urbanization of Capital.* Baltimore, MD: Johns Hopkins University Press.

—— (1985b) 'The geopolitics of capitalism', in D. Gregory and J. Urry (eds.) *Social Relations and Spatial Structures.* New York: St. Martin's Press. 128–63.

Johnston, R. J. (1979a) *Political, Electoral and Spatial Systems.* Oxford: Clarendon Press.

—— (1979b) 'Class conflict and electoral geography', *Antipode* 11,3: 36–43.

—— (1980) 'Electoral geography and political geography', *Australian Geographical Studies* 18: 37–50.

—— (1982) *Geography and the State: an Essay in Political Geography.* London: Macmillan.

—— (1984b) 'The political geography of electoral geography', in P. J. Taylor and J. House (eds.) *Political Geography: Recent Advances and Future Directions.* London: Croom Helm, 133–48.

—— (1985) *The Geography of English Politics: the 1983 General Election*. London: Croom Helm.

—— (1986a) 'Placing Politics', *Political Geography Quarterly*, Supplement to 5: S63–S78.

—— (1986b) 'Places and votes: the role of location in the creation of political attitudes', *Urban Geography* 7,2: 103–17.

—— (1986c) 'Places, campaigns and votes', *Political Geography Quarterly*, Supplement to 5: S105–S117.

—— (1986d) 'A space for place (or a place for space) in British psephology', *Environment and Planning A* 19: 599–618.

—— (1986e) 'The neighbourhood effect revisited, *Environment and Planning D: Society and Space* 4: 41–55.

—— (1987a) 'The geography of the working class and the geography of the labour vote in England, 1983', *Political Geography Quarterly* 6: 7–16.

—— (1987b) 'Dealignment, volatility, and electoral geography', *Studies in Comparative International Development* 22: 3—25.

Key, V.O. (1949) *Southern Politics in State and Nation*. New York: Knopf.

McAllister, I. and Rose, R. (1984) *The Nationwide Competition for Votes: the 1983 British Election*. London: Frances Pinter.

Murauskas, G. T., Archer, J. C. and Shelley, F. M. (1988) 'Metropolitan, non-metropolitan and sectional variations in voting behavior in recent presidential elections', *Western Political Quarterly*, forthcoming.

Osei-Kwame, P. and Taylor, P. J. (1984) 'A politics of failure: the political geography of Ghanaian elections, 1954–1979', *Annals of the Association of American Geographers* 74: 574–89.

Reynolds, D. R. and Knight, D. B. (1989) 'Political geography: recent practices and future relevance', in G. L. Gaile, and C. J. Willmott (eds.). *Geography in America*, Chicago: Merrill.

Sanguin, A.-L (1985) 'Political geographers of the past II: André Siegfried, an unconventional French political geographer', *Political Geography Quarterly* 4: 79–83.

Savage, M. (1987) 'Understanding political alignments in contemporary Britain: do localities matter?', *Political Geography Quarterly* 6: 53–76.

Sayer, A. (1984) *Method in Social Science: A Realist Approach*. London: Hutchinson.

Schattschneider, E. E. (1960) *The Semi-Sovereign People*. New York: Holt, Rinehart & Winston.

Shelley, F. M. (1988) 'Structure, stability and section in American politics', *Political Geography Quarterly* 7: 153–60.

Taylor, P. J. (1978) 'Progress report: political geography', *Progress in Human Geography* 2: 153–62.

—— (1981) 'Political geography and the world-economy', in A. D. Burnett and P. J. Taylor (eds.) *Political Studies from Spatial Perspectives*. Chichester: Wiley, U.K., 157–74.

——(ed.) (1982) 'Editorial essay: political geography – research agendas for the nineteen eighties', *Political Geography Quarterly* 1: 1–17.

—— (1984a) 'Accumulation, legitimation and the electoral geographies within liberal democracy', in P.J. Taylor and J. House (eds.) *Political Geography: Recent Advances and Future Directions*. London: Croom

Helm, 117–32.

—— (1984b) 'The geography of elections', in M. Pacione (ed.) *Progress in Human Geography*. London: Croom Helm.

—— (1985) *Political Geography: World-Economy, Nation-State, and Locality*. New York: Longman.

—— (1986) 'An exploration into world-systems analysis of political parties', *Political Geography Quarterly*, Supplement to 5: S5–S20.

—— and Johnston, R.J. (1979) *The Geography of Elections*. Harmondsworth: Penguin.

Thrift, N. (1983) 'On the determination of social action in space and time', *Environment and Planning D: Society and Space* 1: 23–58.

Wallerstein, I. (1979) *The Capitalist World-Economy*. Cambridge: Cambridge University Press.

Part II

The cleavage model and electoral geography

Chapter four

The electoral geography of the Netherlands in the era of mass politics, 1888–1986

N. P. Passchier and H. H. van der Wusten

On 6 March 1888 the Netherlands witnessed its first election under a new constitution. The electoral system and the franchise had been changed after extensive debate. Membership of the Second Chamber had been set at 100, normally elected in single member constituencies, and the number of eligible voters had more than doubled.

The election results were remarkable. For the first time a confessional majority was returned and the first Social Democrat entered parliament. Looking back, this election marks the beginning of a new era. Modern political mass parties with a stable, organized relationship between electors and elected started to dominate parliamentary politics, replacing the loosely connected clubs of liberal and conservative notables.

In this chapter we shall explore the electoral evolution of the main political currents in the Netherlands since 1888 and reconsider the question posed by Johnston *et al.* (1983: 185) in a similar context: '... whether the changes in the social cleavages have created new electoral geographies'. Their paper was concerned with stability and change in Dutch voting behaviour since the Second World War as derived from aggregate data. It attempted to indicate the consequences of changes among the electorate for the shifts in the geographical patterns of voting. To identify the main political 'tendencies' factor analyses were used correlating voting results over time for spatial units, a method previously applied by Archer and Taylor (1981) in their study of the US presidential elections.

Our aim and method are similar. However, we considerably extend the time span of the study. Johnston *et al.* (1983) started from the electoral patterns in the Netherlands in the 1940s and 1950s, an extremely stable situation which they contrasted with the changes that occurred from the 1960s. In the present chapter we shall attempt to demonstrate the historical background of this stable pattern. It was in fact the result of a long process whereby the modern Dutch political parties became established as mass organizations, a process that started around 1888. Our aim is therefore not only to demonstrate the changing electoral geography since the 1960s after the mould of Dutch politics had been broken but also to portray the electoral geography of the pre-1920 period, before the Dutch party political pattern became largely frozen, like European party systems generally (Lipset and Rokkan, 1967).

Electoral reform

After the extension of the franchise in 1888 about a quarter of the Dutch adult male population had the right to vote. During the following decades franchise-extension was continuously an issue in parliamentary politics. It was one of the main causes of splits within parties in the 1890-1910 period. In 1894 the government was defeated over a Franchise Bill that came close to universal male suffrage. In the new parliament the opposing majority passed a less extreme Bill, limiting voting rights to about 47 per cent of the male population over 25 years of age. However, the possibilities of qualifying as a voter were extended to such a degree that this percentage increased to 68 in 1913 (Stoppelenburg and Van der Wusten, 1986). This has been one of the factors enabling the Social Democrats to grow strongly. Their agitation focused on the franchise as can be seen from a number of massive rallies and a widely endorsed petition campaign in 1912 (Van der Meer *et al.*, 1981).

During the First World War universal male suffrage was introduced. This decision was part of a package deal between the main parties, including the Social Democrats. It resolved a number of divisive issues from the past decades. Lijphart (1975) has pointed to this Pacification of 1917 as the basis of Dutch politics for the next generations, the era of accommodation.

With regard to male suffrage the electoral reforms of this period can be seen as the final, hardly surprising, result of an old debate at a time when 70 per cent of all adult males had already acquired voting rights under the existing law. Other parts of these reforms meant a sharper break with the past. First, women got voting rights after the war. This produced many more new voters on the register in 1922 than the male franchise extension had done in 1918. Second, voting, or rather appearance at the ballot box, became compulsory. This strongly heightened electoral participation, notably in the cities and in the Catholic regions in the south. Third, and probably most important, proportional representation was introduced. Local representation was abandoned for a list system in which seats became allocated on the basis of a national quota.

Since then Dutch electoral law has been changed only in a piecemeal fashion. The abolition of compulsory voting in 1971 has been the exception. Initially this induced a sharp decrease in turnout (Irwin, 1974). Later, turnout at parliamentary elections stabilized around 80 per cent, which is fairly high in a comparative perspective. The rather 'extreme' electoral system is, however, rarely disputed (cf. Daalder, 1975). The Dutch electorate and the politicians seem to cherish their system.

The party system

Proportional representation has significantly affected the party system. The fragmentation of parliament has not been the most important consequence, as one would assume following Hermens, Duverger and others. Indeed, as the threshold is very low, small political parties have often been able to win seats. Only rarely, however, have these small parties secured more than 15 per cent. Further, most small parties are relatively stable. They perceive their position within the political system as an ideological corrective to a larger party of the same family. Only some of them can be considered as 'anti-system flash parties'.

Paradoxically, the most important consequence of proportional representation has been a strengthening of the position of the main parties. In spite of all sorts of fissions and fusions the Dutch party system shows much continuity: the current system parties are the direct successors of those involved in the Pacification deals of 1917 as the Liberals, Anti-Revolutionaries, Christian-Historicals, Catholics and Social Democrats. All survived the Great Depression and the following world war. Increasing secularization during the 1960s and the 1970s forced the big confessional parties to merge into one Christian Democratic party in order to maintain their power position, but still the similarities of the contemporary political landscape and that of three generations ago are striking.

Daalder (1975: 198) has emphasized this continuity. In spite of certain shifts in electoral preferences, the Dutch party system is and has always been characterized by multi-party pluralism. Even in 1888–1913, when the Netherlands moved in the direction of Anglo-Saxon adversary politics, clear two-party polarization never occurred. Proportional representation was a consequence more than a cause of the pluralist party system. It was an attempt at a fair deal. In the ensuing Era of Accommodation, proportionality became a general organizing principle of public life.

In retrospect the Dutch party system of the last hundred years at first sight shows a huge number of groups and groupuscules. This applies long before 1918.

The Liberals started a League in 1886, but they had great difficulty transforming it into a united party. Efforts in that direction failed for a long time. Mostly a right and a left liberal party remained in existence, as in Germany and France. Only after 1945, when Liberalism had reached its lowest ebb, was a united party successfully launched. However, during the re-emergence of Liberalism in the 1960s and the 1970s a new party called Democraten '66 (D'66) arose, ideologically embodying the tradition of the left Liberals.

The confessionals had Catholic and Protestant parliamentary parties even before 1888. Although they made electoral pacts in constituencies where both were in a minority position (Stoppelenburg and Van der Wusten, 1986: 244), cooperation was not easy. The Protestant Anti-Revolutionary Party had to cope with splits in various directions even before 1918 and orthodox Calvinist mini-parties have occupied seats in parliament through to the present day. Catholics have known similar splits all through the period, but none of these existed for long.

Diversity among socialists developed as in other European countries. The mainstream Social Democratic Workers Party (SDAP) started in 1894 after the failure of a previous league. A leftist offshoot, established in 1907, eventually became the Communist Party. This in its turn has produced an intricate history of splits and exclusions ever since the 1920s. During all these years there have also been fragile non-communist parties to the left of the Social Democrats. On the other side the Social Democrats merged in 1946 with parts of the Liberal and Christian left into the Labour Party (PvdA). In 1972 a rightist split of this merger, called Democratic-Socialists (DS'70), was successful for a very short time but has since disappeared.

In spite of this multitude of organizations a considerable degree of coherence and continuity is to be discerned. Three blocs had already crystallized by 1918: liberals, confessionals and socialists. Within these blocs a few parties have held predominant positions. They were potential coalition partners after 1918 and may be called 'system-parties'. Within the liberal bloc these were the Liberale Unie, later the Liberale Staatspartij, after 1945 the VVD. Among the confessionals these have been the Anti-Revolutionaire Partij, the Christelijk-Historische Unie and the Catholics, recently merged into the Christian Democrats (CDA). Within the socialist bloc the SDAP and its successor the PvdA have been central. These main system parties have collectively dominated parliamentary politics. On average they have polled about 85 per cent of the valid vote. Most of the smaller parties have an ideological position alongside and at the same time directed towards one of these system parties.

As a consequence we assume that the principal political cleavages over a period of a hundred years are not so much between individual parties as between 'familles politiques'. This point of view is by no means new, having been put forward in one of the first essays on the electoral geography of the Netherlands (Litten, 1936).

Of course, the major operational problem is the attribution of parties to the various political tendencies. Some of the smaller parties have been situated more or less at the edge between two blocs. Therefore, they have been considered as a separate category at earlier occasions (see e.g. Johnston *et al.*, 1983, who call them 'bridging parties'). It remains to be seen whether the political blocs will then be better delimited. We have preferred as much as possible to classify parties according to the position that party leadership has chosen, e.g. by selecting a name tag, by attempting to merge with other parties, by announcing adherence to international organizations. Marginal differences of opinion on the matter are, however, bound to remain.

For the present analysis a distinction has been drawn between Liberals (including D'66), Socialists (including DS'70), Protestants, Catholics (the last two also taken together as Confessionals) and a remaining category of Others. The latter are often rightist anti-system parties, very difficult to classify from an ideological point of view. The best labels would perhaps be Ultra-nationalist or Ultra-conservative. Some other parties in this category concentrate on the articulation of specific interests, e.g. shopkeepers or farmers. All share a position outside the party system just elaborated and can be characterized as 'flash parties'. There are also 'anti-system parties' of the left, but these tend to participate emphatically in the day-to-day politics of parliamentary democracy and they have been classified as Socialists. The rightist anti-system parties mostly lack this type of 'negative integration'.

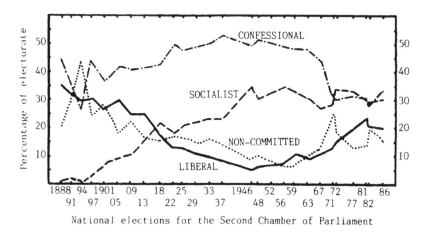

Figure 4.1: Electoral development of the political tendencies, 1888–1986

For this reason we have in Fig. 4.1 added the votes for the Other parties to the abstentions to form a category of Non-committed (to the principles and practice of the current political system that is). Thus we express the idea that non-commitment is an option besides the preference for one of the larger or smaller ideological parties. Depending on the rules (e.g. compulsory voting or not) and the range of party choices offered, this option can be used in a number of ways (abstention, blanc, support outsiders). In practice abstention has been the normal way and in Figure 4.1 most of the Non-committed have in fact abstained. That is why this category was so small under the rule of compulsory voting.

It is interesting to see the clear relation over time between the percentage of Non-committed and the Liberals. This may reflect social and political developments in the Era of Accommodation. In the pillarized society between the 1920s and the 1960s individuals within the confessional and socialist blocs were maximally tied to their various organizations. In that type of society Liberals were to a certain extent outsiders as well. The politics of that period were, following V. O. Key's phrase, 'group politics'. Through their multiple links via mass organizations with strong local chapters, Confessional and Socialist parties were able to mobilize their voters to a maximal degree. It was only in the Era of Dealignment that Liberals got new chances. From this perspective Fig. 4.1 shows three historical periods:

I. Pre-1918, the period of franchise extensions. The increase in the number of voters among the poor was advantageous for the Socialists and to some extent also for the Confessionals. As a consequence Liberal support declined. Election campaigns became more and more important in this period. We call this period the Mobilization Era of the party system.

II. 1918–63, the period of pillarized society. The Confessionals and the Socialists dominated electoral politics at the expense of the Liberals. After the Second World War these pillarized blocs began to stagnate but this did not yet lead to any major shifts in electoral preferences. We call this the Accommodation Era after Lijphart (1975).

III. Post-1966, the period of volatility in Dutch politics and its consequences. Religious conflict, particularly among Catholics, resulted in rapid secularization and a sudden electoral decline of the Confessional parties. However, the restlessness has been more general. As in other European countries, the Dutch electorate has become more volatile since the 1960s. This has been enhanced by the abolition of compulsory voting and the growing number of young voters. The Liberals have largely profited from these waves of unrest. We call this period the Dealignment Era.

In this study we want to describe the geographical diversity behind the aggregate national figures of Fig. 4.1. Each of the 27 parliamentary elections since 1888 produced its own geographical distribution of political tendencies. What continuities and changes can be discerned in these patterns? Johnston *et al.* (1983) have demonstrated the emergence of a new pattern in the 1970s which gradually modified the electoral geography of the preceding period. We presume that three distinctive electoral geographies of the Netherlands have existed during the whole period that mass politics was in operation, connected with the three eras that we have indicated.

Data and methods

To study the development of the electoral geography over the past century it is necessary to construct data matrices with comparable units of observation and comparable political tendencies across 27 elections. As a consequence of the length of the period of study this is by no means easy to do. We have indicated the changes of political organizations in the preceding section. They caused problems as did the change in the electoral system in 1918 which strongly altered the spatial units used to aggregate votes.

The delineation of some historically defensible political tendencies has enabled us to make political preferences comparable over time. On account of the originally very sharp social cleavage between Catholics and Protestants we have classified the political parties of these sub-cultures separately until the final merger. We also took them together as Confessionals for the whole period. A separate category of Non-committed has been introduced, similar to Fig. 4.1. This implies that political preferences have been related not to the valid vote but to the total electorate. This is rather unusual and deviates from the way in which Johnston *et al.* (1983) proceeded. As we consider abstention and spoiling votes to be real options for electors, we argue that preferences as a percentage of the total electorate give a better idea of the relative attractions of the various tendencies.

The selection of units of observation has given much trouble because of the varying quality of data sources. Electoral statistics at the level of municipalities have been published only since 1933. For the 1918–29 period official results are available only at the level of 'Kamerkies-kringen' (enumeration units derived from provinces that hardly enable rural–urban differentiation because of their large size: see Johnston *et al.*, 1983: 189). With difficulty we could reconstruct election results for a number of municipalities from press reports. For the 1897–1913 period we could rely on published statistics for the 100 constituencies. The data for 1888–94 had to be derived from less reliable sources, which ultimately again meant press reports. It proved impossible to connect the earlier constituencies precisely with the later 'Kamerkieskringen'. Finally we had to be satisfied with relevant units that had a maximal overlap for the pre- and post-1918 period (14 major urban areas and 15 rural areas in surrounding 'Kamerkieskringen'). The population overlap is always 90–100 per cent, with one exception of 70 per cent.

In summary our initial data base splits the electorate into supporters of Liberal, Socialist, Catholic and Protestant candidates or lists, plus the Non-committed. This split has been done for 29 units of observation over 27 elections.

To derive geographical patterns in their development over time from these data we have used the same method as Johnston *et al.* (1983). In that respect our study replicates their work. In addition we provide more geographical detail (29 instead of 18 units) and more historical depth (27 instead of 11 elections).

We constructed six data matrices, one for each of the four tendencies plus one for the Confessionals as a whole and one for the Non-committed. They contained the percentages of the total electorate supporting the tendency in each of the 29 units. The elections were the variables. These were correlated producing six matrices that indicated the similarities of the geographical patterns in the various elections. Factor analysis on these matrices used SPSS principal factor extraction with oblique rotation and delta set at 0.0 (factors may therefore correlate fairly highly). This choice has been argued in Johnston *et al.* (1983: 190). We deviated at one point from their procedure. The minimal eigen value was set at 1.0, contrary to the unusually low 0.3 in the earlier study, selected by Johnston *et al.* on the basis of experiment in order to produce second factors. In our factor analyses various factors appeared in most cases before the eigen value cut-off point of 1.0 had been reached. That is why we followed standard practice to allow only factors that contributed more to the variance than one variable.

Results: periods

In accordance with our hypotheses the six factor analyses produced dimensions related to the historical periods previously described. The changes in the electoral geography of the Netherlands roughly followed the transitions of 1918 and 1967.

Evidence for this conclusion is shown in Table 4.1. In the last column factor pattern loadings are used to indicate which elections have been primarily determined by the various factors. These pattern loadings measure the direct relationship between a variable and a factor, independent of the effects of other factors. As they clearly assign distinct factors to the variables, i.e. elections, these coefficients are the most suitable for historical periodization. Indeed, in most cases the factors discerned reflect periods, i.e. successive elections have similar loadings.

Table 4.1: Summary of factor analyses

Factor	Eigen values	Percentage of variance		Inter-factor correlations			Elections with pattern loadings
		Factor	Cum	F1	F2	F3	over 0.5
Liberal							
1	18.2	67.5	67.5				1888–1963
2	4.9	18.2	85.5	0.35			1967–86
3	1.4	5.3	91.0	-0.11	-0.09		1918
Socialist							
1	20.1	74.4	74.4	0.50			1918–72
2	2.9	10.9	85.3	0.19	0.37		1897–1913
3	1.4	5.1	90.3	0.63	0.42	0.18	1891–4
4	1.0	3.8	94.1				1977–86
Catholic*							
1	21.5	93.4	93.4				1888–1972
Protestant*							
1	20.1	87.6	87.6				1888, 1918–72
2	1.4	6.1	93.6	0.74			1891, 1897–1913
All confessional							
1	22.5	83.4	83.4				1897, 1905–67
2	2.3	8.7	92.1	0.70			1971–86
3	1.0	3.9	95.9	0.59	0.34		1891–1901
Non-committed							
1	10.7	39.8	39.8				1888–1913, 1967
2	7.4	27.3	67.1	0.01			1918–37, 1948–63
3	2.7	10.0	77.2	-0.35	-0.21		1918, 1971–86

* Only 23 elections up to 1972. The other analyses included 27 elections.

A further clarification of the factors is provided in Fig. 4.2, showing the factor structure loadings. These are the overall correlations between factors and variables, measuring the total relationship and reflecting both direct and indirect effects. The latter are important in this study because of the sizeable intercorrelations between factors as a result of oblique rotation. In a number of cases more elections turned out to be correlated to the factors than was to be expected from the pattern

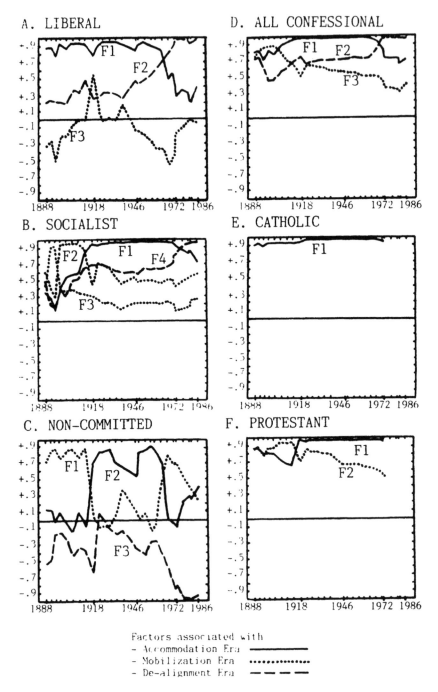

Figure 4.2: The correlation between factors and variables (factor structure)

loadings. Obviously, there have been long-term influences in the electoral geography of the Netherlands beside the peculiarities of individual historical periods.

The electoral geography of the Non-committed went through the sharpest contrasts during the three historical periods. The curves of the factor structures (Fig. 4.2c) are clearly shifting, crossing over in 1918 and 1967. The first dimension shows maximal loadings in 1888–1913 and again in 1967–72. In contrast, the second is positively related to the Accommodation Era when turnout was much higher. Institutional changes in the political system around 1918 and 1967, such as the adoption and abolition of compulsory voting, may explain the alternate importance of the two patterns. The third factor, with its maximal loadings on the elections since 1970, seems to reveal the modern geography of electoral participation following the abolition of compulsory voting.

The factor analysis of the political tendencies all produced first factors associated with the Accommodation Era. The percentage of variance explained by these first factors varies between 68 per cent and 93 per cent (Table 4.1), much more than the 48 per cent expected from the fact that 13 of the 27 elections fell between 1918 and 1967. High loadings on other elections, before and after the Accommodation period, account for this figure (Fig. 4a-b, d-f).

The traditional pattern of the Liberal tendency appears to be older than 1888; after 1963 its relevance quickly diminished. The reverse holds for the Socialists. Their dominant geographical pattern only developed during the Mobilization Era. Since then, however, it has remained important, still correlating highly with the Socialist vote of the 1980s. The distribution of the Confessional tendencies, Catholics as well as Protestants, has been most stable during the past century. Their first factors, although showing pronounced loadings during the Accommodation Era, correlate strongly with elections both before 1918 and after 1967.

Consequently, we may conceive these first factors as the aggregation for each of the political tendencies of the general pattern and the specific influence of the Accommodation period. The other factors may be understood as time-specific deviations from these patterns during the Mobilization and Realignment periods.

Specific voting patterns for the Mobilization Era are most clearly discernible for the Socialists, with separate factors describing early and later penetration of the electorate. The Confessional vote, the Protestant vote in particular, has also been affected by particular influences. In Fig. 4.2 the distinctive character of the geography of voting in that early period is documented by the cross-over of Socialist and Confessional factors around 1910. However, these patterns are strongly correlated to

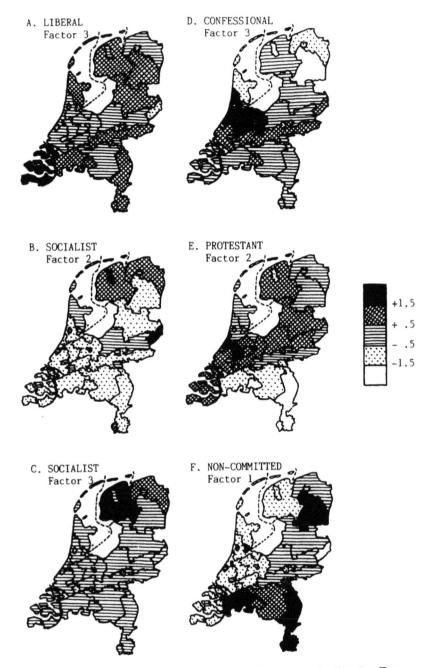

Figure 4.3: Factor scores on factors asociated with the Mobilization Era (pre-1918)

the general distributions of Socialist and religious voting. Liberal voting patterns hardly changed before the First World War, probably because initial Liberal mobilization had occurred before 1888.

As to the Dealignment Era, the results of Johnston *et al.* (1983) are largely confirmed. All main tendencies show a cross-over around 1970. Since then a new geographical pattern of voting has emerged, which is not totally independent of the general distribution of the tendencies however. In the analyses of Socialists and Confessionals the dealignment dimensions are strongly correlated with earlier elections.

Results: factors

From the preceding section we conclude that it is proper to distinguish three separate electoral geographies in the Netherlands. Each of the historical periods, the Mobilization, Accommodation and Dealignment Eras, seems to be characterized by a distinct voting pattern. In the following discussion of factor scores an attempt has been made to describe these varying regional patterns (Figs. 4.3–5). In classifying the factors by period we focus our attention on historical developments. In studying factor scores we try to establish geographical relations.

Mobilization Era (Fig. 4.3)

The *Socialists* were new entrants in the electoral arena of the period. The factor analysis for this tendency has two dimensions with maximal loadings before 1918. *Socialist factor 3*: Independent Socialism (Fig. 4.3c). This factor takes only 5 per cent of the variance: its greatest strength is in 1891–94. There is one extreme score in rural Friesland, and there are some lower ones in the Groningen countryside and in the cities of Arnhem and Utrecht. To a large extent these may be explained by the prehistory of the SDAP. In 1891 its predecessor, the SDB, had fallen apart after disputes concerning parliamentary politics and electoral participation. In most constituencies no socialist candidates were put forward. Only in the north, e.g. in Friesland where Domela Nieuwenhuis had won the first socialist seat in 1888, were a number of independent socialist candidates campaigning successfully under various labels.

Socialist factor 2: Organized Socialist Mobilization (Fig. 4.3b). This factor explains 11 per cent with high loadings for 1897–1913. Scores are high in a number of peripheral towns and again in the northern countryside. Middle range scores, but still above average, are to be found in the western cities and rural North Holland. Here the SDAP actively mobilized in its formative years in strong competition with the Liberals. Most new seats in the breakthrough election of 1913 were gained in these areas, which can properly be called early red strongholds, where the Social Democrats have dominated local government up to the present. This entrenchment is reflected in the correlation of this factor with the present pattern of socialist support (still $r = 0.52$ after 80 years).

The *Confessionals* also have two factors with high loadings before 1918.

Confessional factor 3: Early Interconfessional Contest (Fig. 4.3d). This dimension only explains 4 per cent: the intercorrelation with Confessional Factor 1 is high ($r = 0.59$). Consequently the factor loads on nearly all elections, but with an emphasis on 1888–1905. Factor scores are high in religiously mixed zones in the west, where Catholics and Orthodox Protestants together held a majority (Stoppelenburg and Van der Wusten 1986: 443). In those early years confessional electoral co-operation had not yet developed and candidates for both tendencies were regularly put up, thus increasing turnout. Because of confessional electoral pacts these seats became less contested in 1909 and 1913 (Verhoef, 1974) and turnout diminished.

Middle-range scores refer to one-tendency majority regions, where Protestants and Catholics never put up opposing candidates. Lowest scores indicate areas where confessionals had poor chances and thus conducted tepid campaigns only.

Protestant factor 2: Early Confessional Cooperation (Fig. 4.3e). This dimension correlates highly with Protestant factor 1 ($r = 0.74$). Loadings are highest in 1905–13. This factor reflects the areas where Protestant candidates got more votes than expected from the size of their own constituency (Stoppelenburg, 1984: 48–9). In most cases Catholics supported Protestant candidates as a consequence of the electoral pacts just mentioned.

Liberals had been active in Dutch politics long before the start of the Mobilization Era. Their principal support pattern was already in place and hardly changed until the Dealignment Era. The interpretation of the Liberal factor in Fig. 4.3, that is perhaps most characteristic for the first period, is difficult.

Liberal factor 3: Liberal Urban–Rural Differences (Fig. 4.3a). This factor explains 5 per cent and loads negatively on 1894 and 1967, positively on 1918. Factor scores are positive in peripheral rural regions, they are lower or negative in cities. This is difficult to explain. It comes down to a better showing of the Liberals in towns and cities in 1894 and 1967 and the reverse in 1918: more losses in urban areas. This may be related to a sharpening of internal conflicts among Liberals around those years with varying reactions among liberal voters in town and countryside.

The distribution of *Non-committed* during the Mobilization Era appears to be related to institutional characteristics of the electoral system and to tactical voting.

Non-committed factor 1: Early Safe Seat Apathy (Fig. 4.3f). This factor explains no less than 40 per cent with high loadings in 1888–1913. Factor scores are particularly elevated in the southern provinces and in Drenthe, the undisputed strongholds of Catholics and Liberals respectively. The labelling of this factor particularly refers to these cases. High scores in western cities, however, have nothing to do with unopposed elections. The factor shows high loadings again in the 1970s, after the abolition of compulsory voting. Likewise, safe seat apathy cannot account for these. It has been hypothesized (Passchier, 1988) that political culture, particularly in the southern part of the country, has localist orientations that may have induced low participation in national elections during the 1970s.

Accommodation Era (Fig. 4.4)

In almost every factor analysis the most important dimension was associated with this period. The conclusion earlier was that these first factors may be considered as the aggregations of general patterns and specific influences of the period. Some of these factors are clearly related to the religious map of the Netherlands: Catholics in the south, an orthodox-Protestant ribbon from Friesland in the north to Zeeland in the south-west, latitudinarian Protestantism in many towns and in the rural parts of Groningen, North Holland and various areas in Gelderland (Daalder, 1981: 204 ff., map in Stoppelenburg and Van der Wusten, 1986: 443). It is therefore appropriate to begin the description of the pattern of this period with the distribution of the various types of *Confessional* voters.

Catholic factor 1: Catholic Residence (Fig. 4.4e). This dimension (93 per cent) reflects the religious map of Catholicism in the Netherlands before 1970 almost perfectly. A Catholic majority in the south, Catholic minorities in the west and in Overijssel, a Catholic diaspora in the north.

Protestant factor 1: Protestant Residence (Fig.4.4f). This dimension (88 per cent) reflects the orthodox Protestant ribbon. As in the case of the Catholics but more so, towns and cities have lower scores than their hinterland.

Confessional factor 1: Confessional Resistance to Secularization (Fig. 4.4d). This factor (83 per cent) takes the Protestant and Catholic patterns together. Protestant areas are in a political sense less homogeneous than Catholic ones, where the highest peaks of political confessionalism before 1967 are to be found. Areas with widespread latitudinarian support – and these are the same regions where

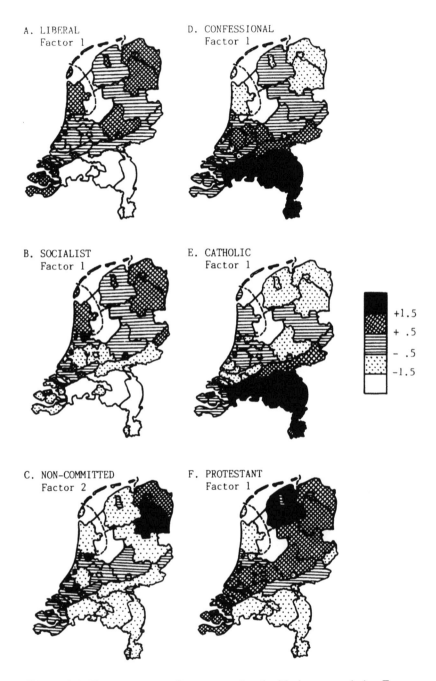

Figure 4.4: Factor scores on factors associated with Accommodation Era (c 1918–63)

secularization started early on (Kruijt, 1933) – are clearly marked in a negative sense: the cities, North Holland, Groningen, Drenthe.

Socialist factor 1: Socialist Secularization and Urbanism (Fig. 4.4b). This factor (74 per cent) has a very strong negative correlation ($r = -0.97$) with Confessional factor 1, which is particularly telling given that they have been produced by two different factor analyses. Rural–urban differences show up strikingly. In pillarized society the Confessionals had their centre of gravity in rural and small town settings whereas Socialism was largely confined to the larger cities. Originally this had not been the case. The Socialists started to mobilize in the north (cf. Socialist factors 1 and 2). Between 1905 and 1913 they 'conquered' the western cities, where industrialization had definitely taken off and class conflict offered opportunities to mobilize. At the same time the threshold to enter the 'pays legal' was lowered. In the industrializing areas in the south, more rural and overwhelmingly Catholic, their attempts to penetrate were unsuccessful. As a consequence the growth of Socialist support stagnated after 1918 and the electoral centre of gravity remained situated in Protestant latitudinarian areas and particularly in cities.

Liberal factor 1: Liberal Latitudinarianism (Fig. 4.4a). This factor (68 per cent) also has a very strong negative correlation ($r = -0.89$) with Confessional factor 1. Rural–urban differences show up again but are less pronounced than for the Socialists. Liberal and Confessional factors 1 load highly on the pre-1900 elections. Apparently the political relevance of the religious map of the Netherlands had been firmly established by that time. The electoral geography of the Liberals has from the dawn of mass politics up to the 1970s primarily been determined by the cleavage about church–state relations that had come to the fore in the third quarter of the nineteenth century.

Non-committed factor 2: Outsiders of Pillarized Society (Fig. 4.4c). This factor (27 per cent) is also negatively correlated with Confessional factor 1 ($r = 0.80$). The geographical pattern is quite different from the pattern of Non-committed factor 1. Particularly in the confessional majority areas in the south and the east proportional representation and compulsory voting increased electoral participation. 'Safe seat apathy' disappeared, for after 1918 each vote counted, from whatever part of the country it originated. The confessionals were able to collect all the potential votes on account of their dense network of organizations in all spheres of life. The Liberals adapted with far more difficulty. In their traditional stronghold Drenthe, an electorate developed with only restricted ties to national politics during the inter-war period. A similar story can be told for a sizeable part of the urban middle classes. During the 1930s these non-integrated outsiders of pillarized society largely determined the electoral opportunities of the Dutch national–socialist party (Passchier and Van der Wusten, 1979).

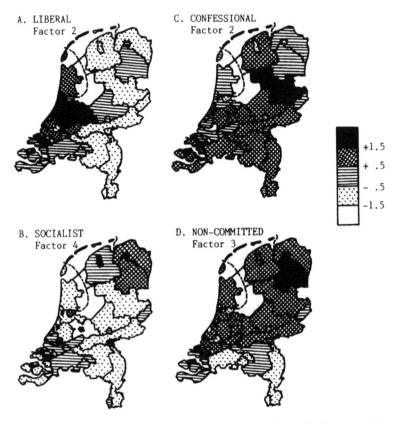

Figure 4.5: Factor scores on factors associated with the Dealignment Era (after 1963)

Dealignment Era (Fig. 4.5)

This new period and its novel electoral geography have already been described by Johnston *et al.* (1983). We limit our discussion to a comparison with their results.

Confessional factor 2: Confessional Rural Resilience (Fig. 4.5c). This factor also refers to a similar one in Johnston *et al.* (1983). It is strongly correlated to the principal confessional dimension from the Accommodation Era (r = 0.78) and loads highly on earlier elections. Confessional decrease has occurred largely in the urban areas and in latitudinarian parts of the country, where the process of secularization has strongly progressed after 1970 (Doorn and Bommelje, 1983: 33).

Socialist factor 4: Socialist Strongholds (Fig. 4.5b). This is also the counterpart of a factor identified by Johnston *et al.* (1983). Like the Confessional factor for the Dealignment Era it is strongly correlated to the factor found for the Accommodation Era ($r = 0.63$). It is therefore not surprising that it explains only a mere 4 per cent of Socialist variation. Highest factor scores are found in the traditional areas of Socialist mobilization: the north-east and the cities. The Socialist tendency appears in recent years to have become even more urban than it already was.

Liberal factor 2: Modern Liberal Suburbanism (Fig. 4.5a). This factor (18 per cent) is almost identical to the one Johnston *et al.* found. As we studied more carefully delimited areal units the suburban character of the recent Liberal advance in the western provinces is more clearly displayed.

Non-committed factor 3: Modern Democratic Reluctance (Fig. 4.5d). This factor (10 per cent) is largely independent of other dimensions. Two trends predominate: a north–south difference and an urban–rural split (Passchier, 1988). In the northern and eastern countryside system parties have been able to stabilize their support, in the cities and the Catholic majority areas sizeable portions of the electorate have distanced themselves from national elections together.

Conclusion

Three periods can be observed in the evolution of the electoral geography of the Netherlands since the beginnings of mass politics. The geographical relations of one of these, the Era of Accommodation, have been most influential. Indeed, the political map of pillarized society has dominated the past century. However, in the Mobilization era, as well as in the Dealignment Era, distinctive voting patterns occurred, modifying this general map.

The electoral geography of the Accommodation Era has been determined primarily by the divide between confessionals and non-confessionals. The territorial map of the religious divisions in the Netherlands has largely shaped political support patterns. This territorial pattern of religious adherence had been established in an earlier part of the nineteenth century. Secularization in the twentieth century has slightly altered its features, particularly by sharpening the rural–urban split in religious attachments and consequently in political preferences.

In the electoral geography of the Mobilization Era the main pattern determined by religion has been modified in two ways. First, class conflict became increasingly salient after 1888, when industrialization was still in its infancy. Its earliest manifestations were to be found in peripheral rural areas (e.g. peat colonies) and in old industrial towns (e.g. Maastricht), later in the booming towns of modern industry (e.g. Enschede in the east), and finally the big western cities. The mobilization of Socialists follows this pattern. Secondly, institutional factors had an impact. The electoral system had a definite influence on confessional support patterns during that period. Our analyses showed the effects of safe seat apathy, intra-confessional competition, and inter-confessional electoral co-operation.

The electoral geography of the present Dealignment Era remains strongly influenced by the pattern of the preceding decades. This is astonishing when we realize that the direct influence of church and religion on social life has to a large extent disappeared. Dutch society is no longer different from the most secularized countries of the Western world. Yet the attraction of the political tendencies largely follows the old territorial cleavage of religion and also of urbanization. Confessional decrease occurred primarily in urban areas, while Socialists were better able to maintain their hold there. Large-scale suburbanization processes during the last decades have induced the major modification of the traditional pattern which favoured the Liberals. Along an urban–suburban–rural gradient with these various types differentially represented in the regions, we can rank the relative winners of the present electoral process. They are the Socialists, Liberals and Confessionals in that order (Ostendorf, 1987).

References

Archer, J. C. and Taylor, P. J. (1981) *Section and Party*. Chichester: John Wiley.

Daalder, H. (1975) 'Extreme proportional representation: the Dutch experience' in S. E. Finer (ed.) *Adversary Politics and Electoral Reform*. London: Wigram, 223–48.

—— (1981) 'Consociationalism, center and periphery in the Netherlands', in P. Torsvik (ed.) *Mobilization, Center–Periphery Structures and Nation-building: a Volume in Commemoration of Stein Rokkan*. Bergena and Oslo: 181–240.

Doorn, P. and Bommeljé, Y. (1983) *Maar . . . men moet toch iets wezen*. Utrecht.

Irwin, G. A. (1974) 'Compulsory voting legislation: impact on voter turnout in the Netherlands', *Comparative Political Studies* 7: 292–315.

Johnston, R. J., O'Neill, A. B. and Taylor, P. J. (1983) 'The changing electoral geography of the Netherlands: 1946–1981', *Tijdschrift voor Economische en Sociale Geografie* 74: 185–95.

Kruijt, J. P. (1933) *De onkerkelikheid in Nederland*. Groningen: Noordhoff.

Lijphart, A. (1975) *The Politics of Accommodation*. Berkeley, Cal.: University of California Press.

Lipset, S. M., and Rokkan, S. (1967) 'Cleavage structures, party systems and voter alignments', in S. M. Lipset and S. Rokkan (eds.) *Party Systems and*

Voter Alignments: Cross-National Perspectives. New York: Free Press, 3–64.

Litten, M. (1936) 'De geografische verdeeling der politieke partijen in Nederland voor den wereldoorlog', *Mensch en Maatschappij* 12: 50–9, 126–39.

Meer, T. van der, Schuppen, S. van and Veen, S. (1981) *De SDAP en de kiesrechtstrijd*. Amsterdam: Van Gennep.

Ostendorf, W. (1987) 'Culturele differentiatie binnen Nederland: van territoriale binding naar sociaal-ruimtelijke uitsortering', in H. van der Wusten (ed.) *Postmoderne aardrijkskunde. De sociografische traditie voortgezet*. Muiderberg: 168–79.

Passchier, N. P. (1988) *De geografie van de verkiezingsdeelname*. forthcoming.

—— and Wusten, H. H. van der (1979) 'Het electoraal succes van de NSB in 1935', in F. W. Klein and G. J. Borger (eds.) *De jaren dertig*. Amsterdam, 262–79.

Stoppelenburg, C. M. (1984) 'Het geografisch patroon der politieke verhoudingen in Nederland 1897–1913', Amsterdam (Unpublished Ph.D, thesis).

—— and Wusten, H.H. van der (1986) 'Continuiteit en verandering in de politieke kaart van Nederland 1897–1913', *Acta Politica* 21: 431–48.

Verhoef, J. (1974) 'The rise of national political parties in the Netherlands 1888-1913', *International Journal of Politics*, 207–21.

Chapter five

Tradition contra change: the political geography of Irish referenda, 1937–87

John O'Loughlin and A. J. Parker

Ireland achieved independence in 1922 as a result of a classic peripheral rejection of the British metropole (Lipset and Rokkan, 1967). By the 1950s worsening economic conditions demanded a reversal of the autarchic, nationalist principles on which the independence movement was built before 1920. The economic debate between national idealists and accommodationists is reflected well in Irish referenda over the past decade and a half. The Irish public has made direct statements on Ireland's place in Europe (and by extension in the world economy) and on what kind of social/religious pluralism is acceptable to the majority traditional-Catholic population. The six most important referenda are the subject of this chapter which seeks to understand the changes in Irish political beliefs since the 1936 constitution was approved and to document the geography of the electoral cleavages. These constitutional referenda are not only plebiscites on amendments to the constitution but are statements on how far Ireland has shifted from the 1936 'clerical, nationalist, tradition-bound' world.

In a wider academic context, this chapter attempts to add to the debate on the applicability of the Lipset–Rokkan typology to Ireland. Since Irish party politics are anomalous in Europe (two small left-wing parties with under one-tenth of the vote and three centre-right parties with over 90 per cent), the argument has been made by Carty (1981) that the Lipset–Rokkan model is not applicable in the Irish context. In contrast, Whyte (1974), Garvin (1981) and Sinnott (1984) argue for a modified Lipset–Rokkan cleavage model of Irish politics. By focusing on referenda that concern issues behind Lipset–Rokkan cleavages (economic, national and social choices), the relevance of the Lipset–Rokkan model to Ireland can be further evaluated.

The Lipset–Rokkan model and Irish politics

On the surface, it would appear that all four of the Lipset–Rokkan cleavages are relevant to the Irish situation. Core–periphery, worker-capitalist, church–state and urban–rural disputes are easily identified in Irish history. However, as Sinnott (1984) clarified, the applicability of this cleavage form to Ireland can be validated only if Irish politics are examined prior to independence in 1922 and the post-independence party framework is interpreted in the light of the earlier events. We believe that Carty (1981) is incorrect in suggesting that Irish politics emerged *ex nihilo* in 1922 and it is no surprise that Carty concluded that the four Lipset–Rokkan cleavages are not present in Ireland with that late date as his starting-point. Our interpretation follows that of Whyte (1974), Sinnott (1984), Farrell (1970) and, by extension, the recent updates by Garvin (1987), Mair (1987), Girvin (1986, 1987), Gallagher (1985) and Laver (1986, 1987).

Early in this century Ireland was a classic periphery in both Wallersteinian and Lipset–Rokkan senses. The country was rural, poor, traditional, exploited, and tied intimately to Britain, both politically and economically. A national consensus was formed in 1918 for independence. In the British general election of 1918, the independence movement, Sinn Fein (Ourselves Alone), won 73 of the 105 seats in the island and 70 of the 75 seats in the area that eventually became independent. Although the Labour Party was established in 1912, it stood aside in 1918, subsequently identified as the key mobilizing election, to allow Sinn Fein to present a united electoral front. Farrell (1970) attributed the relative absence of strong socialist support in Ireland (now about 10 per cent of the vote) to this electoral misjudgement; he concluded that since the great mass of voters were not given a socialist alternative (only in Belfast did Labour put up candidates), Labour 'permitted the shaping of a basic cleavage in Irish political life that ran along a constitutional axis and cut across other potential sources of political disagreement' (p. 489). Critical to the argument is the interpretation of the British election of 1918 as the mobilizing election for Ireland. Between 1910 and 1918 the Irish electorate more than doubled and Sinn Fein was the catalyst that formed the enduring electoral alignments. Irish politics became dominated by two versions (idealist and accommodationist) of the nationalist sentiment. The core–periphery cleavage occurred in the wider British context and, as Sinnott (1984: 302) stated, the 1918 general election provided a textbook example of the kind of freezing of party alternatives that Lipset and Rokkan had in mind.

This interpretation begs the question of what happened to cleavages other than the core–periphery division. It is generally agreed that the labour–capitalist cleavage in Ireland was set aside in the face of the nationalist struggle. In fact the Irish Labour Party was strongly nationalist, as exemplified by the writings and actions of James Connolly (Morgan, 1988). The Irish periphery was overwhelmingly

Catholic and had been in bitter conflict with the British government on church–state issues for over 200 years. And as noted by Whyte (1974: 647), the conflict between agrarian and urban industrial interests had been largely subsumed by the wider core–periphery conflict; the net effect was that the Irish Free State after 1922 was homogeneous on three dimensions, being peripheral, agrarian and traditional Catholic.

Secession created the Irish state: the terms of secession created the Irish party system. A split in Sinn Fein over the terms of the treaty led to a 1923 election in the new state that resulted in a contest with 19 parties (O'Leary, 1979). The pro-treaty faction, Cumann na Gaedheal (later the Fine Gael party, League of the Irish), won 63 seats; the anti-treaty faction, Sinn Fein following Eamon de Valera (later the Fianna Fail party), won 44 seats on an abstentionist platform; Labour won 14 seats; a Farmers Party won 15 seats and the rest went to independents. The major cleavage was over the terms of Irish autonomy, between idealist nationalist groups led by de Valera, who rejected the treaty terms that required members of the Dail (parliament) to take the oath of allegiance to the British monarch and to accept that the Governor General represented the king in Ireland, and the accommodationalists who accepted the treaty as a compromise towards a more realistic nationalism. The Cumann na Gaedheal faction anticipated that this gradualism would eventually result in the absorption of Northern Ireland, which had been partitioned from the rest of the island, remaining part of the United Kingdom. In Sinnott's (1984: 303) phrase, the party split was over degrees of commitment to peripheral autonomy. This interpretation of Irish politics places Ireland squarely in the Lipset–Rokkan framework of party development in Europe, accounting for the dominant nationalist consensus and the later split within the independence movement. It also accounts for the homogeneity of Irish politics and the relative unimportance of church–state, rural–urban and capitalist–worker cleavages (Sinnott, 1984).

Given the circumstances of the establishment of the Irish party system, it should not be surprising that ecological analyses of party support in Ireland have failed to yield sizeable coefficients. Both major parties, Fianna Fail (FF) and Fine Gael (FG), are accurately described as 'catch-all' parties with little differentiation of socio-economic or geographical support. O'Leary (1979) showed that class differences in party preference are small by European standards. Using Lijphart's (1971) index of class voting, Ireland's score for 1977 was 6 (the difference in percentages in Labour support among manual and non-manual workers), compared to an average of 50 for Scandinavia and a value of 25 for the other European states (Taylor and Johnston, 1979: 165). FF receives more of the manual worker vote than Labour and it vies with FG for the non-manual worker and farmer votes.

However, Gallagher (1976) was able to show that FF's vote was correlated significantly with localities with a high proportion of (self-identified) Irish-speakers, interpreted as a nationalist commitment, and with a high proportion of small farms. FG's vote correlates with areas of large farms and with high proportions of non-Catholics and skilled employees. Labour has the clearest correlates of all, highest in constituencies with large proportions of urban dwellers (Gallagher, 1976; Parker, 1982, 1986).

When the electoral results are examined at a local scale, below the level of the constituency, the full impact of the personal and local character of Irish elections under the single transferable vote (STV) electoral system becomes apparent. Parker (1982, 1986) has shown that the 'friends and neighbours' effect, measured by distance to the candidates' homes, far surpasses ecological correlates in accounting for the pattern of support. The parties are well aware of this phenomenon and select their candidates with the distribution of local bailiwicks in mind (Sacks, 1970).

Irish government since 1927 has been dominated by Fianna Fail as the only party capable of forming a government without a coalition partner. Typically, FF receives almost 50 per cent of the vote nationally and has been in power for 38 of the 55 years since 1933. The other years had FG-led coalitions with Labour serving in a role similar to that of the Free Democratic Party in West Germany and benefiting from the transfer of the lower preferences of FG voters. The 1980s have seen the appearance of two very different parties, a Workers Party, with its base in poorer urban districts, espousing an uncompromising socialist platform, and a centre-right party, the Progressive Democrats, promoting free enterprise and social–liberal philosophies. In 1987, FF had 44 per cent of the first-preference votes, FG 27 per cent, Labour 6.4 per cent, the Workers Party 4 per cent, the Progressive Democrats 12 per cent, Sinn Fein (the new political wing of the Irish Republican Army, the IRA) received 2 per cent, and 5 per cent went to independents. The two new parties essentially took votes away from Labour (losses to the WP) and FG (losses to the PD) (Parker, 1987). It is still premature to state that Irish politics is either dealigning or realigning (Marsh, 1985). What is clear is that most voters still choose a centre-right party and that the cleavage in the Sinn Fein movement of 1922 dominates the political discussion.

Bunreacht na hEireann (the Irish constitution)

After the electoral victory of Fianna Fail (the anti-treatyites) in 1933 their leader, de Valera, was emboldened to reassert Irish autonomy by the device of a new constitution. It was explicitly designed to reduce the links with the United Kingdom. The constitution, formulated as a personal statement by de Valera in 1936 and approved narrowly (56 per cent voted to accept it) in a 1937 plebiscite, was based on British traditions and the American model, with some special elements that reflected de Valera's views of Ireland's global role and social composition. These were drawn from Catholic social theory and, especially, from papal encyclicals (Chubb, 1970: 67). The document claims Irish sovereignty over the whole island (including Northern Ireland). Articles guaranteeing civil liberties echo the Bill of Rights of the US constitution but Articles 41 to 44 are obviously Catholic in tone and content. Article 44 recognized the 'special position of the Holy Catholic and Roman church as the guardian of the faith professed by the great majority of the citizens'. This article also recognized the other churches and guaranteed freedom of religion; a separate article prohibited divorce. These two articles and a petition to insert an anti-abortion amendment into the constitution were the bases for three referenda, in 1972, 1983 and 1986. Articles establishing Irish sovereignty had to be changed to accommodate Ireland's entry to the European Communities in 1973 and the signing of the Single European Act in 1986. These sovereignty questions were the basis of referenda in 1972 and 1987. With the plebiscite of 1937, these six referenda constitute the data base for this study.

The reason for choosing the Fianna Fail support pattern as the base against which to compare the referendum votes is based on de Valera's personal dominance of this party and his pivotal role in Irish politics. FF's policy is to maintain the traditional aspects of the constitution (Girvin 1986: 80). FF was the party of the periphery which came to dominate the Irish centre in the 1930s and which attempted to codify its view of Ireland (or, more accurately, its founder Eamon de Valera's view) by the 1936 constitution. The plebiscite was viewed at the time as a party political choice, not as a referendum on the state's charter document. Enshrined in the constitution were principles held dear by the strongly nationalist FF group, including protection of the Irish language, support of the family and prohibition of divorce among other Catholic principles, and an irredentist claim to the North. Further, the document enshrined the rural peasant tradition in an article which declared that the state would attempt to 'secure that there may be established on the land in economic security as many families as practicable' and an unusual social welfare–Catholic position that the 'public be protected against unjust exploitation' and that the state 'shall favour and, where necessary, supplement private initiative in industry and commerce'.

Voting in the six referenda can be seen as a measure of how far the electorate has shifted since 1937 and by comparing the FF and referendum votes, we measure how far constituencies deviate from the expected pattern. FF's dominant role in Irish politics and its setting of the national political agenda through party positions on amendment decisions support this procedure. By regressing the amendment votes by constituency with the FF percentage at the closest general election and by examining the residuals from the FF/referenda vote relationships, we identify sub-cleavages as well as highlighting the existing core-periphery split that reappeared in the 1970s. The classification of the 41 constituencies according to referendum voting patterns illustrates the operation of the Lipset–Rokkan model at the local level in Ireland.

There have been 13 referenda in Ireland since 1937 including the plebiscite. Two of the remaining 12 referenda were technical adjustments of a non-controversial character (an adoption Bill 1979 and election of university senators at the same time). Of the others, Sinnott (1988) classifies them as 'regime-related' and 'moral-religious' issues. Five of the seven regime issues govern the rules of elections. Two proposals by Fianna Fail to change the single transferable vote system to a plurality system were defeated in 1959 and 1968. Amendments to extend the vote to 18 year olds in 1972 and to non-nationals in 1984 were passed overwhelmingly (85 and 75 per cent support respectively) and a proposal to allow significant deviation from a proportional allocation of population to Dail deputy was defeated in 1968. The two other regime votes, changing the sovereignty clauses of the constitution to allow entry to the European Economic Community (EEC) and to sign the Single European Act, were both passed by a substantial margin (83 and 70 per cent support: Table 5.1). On the moral–religious referenda, deleting Article 44 was passed in 1972 (84 per cent 'yes'), adding an anti-abortion amendment was passed in 1983 (67 per cent 'yes') and deleting the divorce clause was defeated (63 per cent 'no'). When the issue has been Ireland's relationship to Europe, the changes have been approved, but on all the important domestic issues (changing the electoral system, abortion, divorce) the opponents of change have triumphed (Sinnott, 1988). A final feature of the referenda that is worth commentary is the high proportion of spoilt votes in the 1937 plebiscite (10 per cent) and the 1972 vote on the special position of the Catholic church – Article 44 (5.5 per cent: Table 5.1). These spoilt votes are interpreted as a rejection of the constitutional referendum exercise and/or a preference for something other than the choices offered to the voter.

The changing contexts of constitutional referenda

It is not enough to examine the referenda results in local Irish terms focusing on constituency support levels as the reflection of traditional party loyalties. Party correlates are strong for some referenda but a

Table 5.1: Referendum results in the Republic of Ireland 1937–87

Year	Issue	Turnout (%)			% 'yes'			Spoilt votes %
		Average	Maximum	Minimum	Average	Maximum	Minimum	
1937	Approve constitution	75.8	84.6	61.1	56.5	68.2	36.6	10.0
1972	Accession to EEC	70.9	77.3	62.0	83.1	90.7	72.8	0.8
1972	Delete Article 44 (religion)	50.7	58.6	42.7	84.4	85.4	67.9	5.5
1983	Insert anti-abortion clause	49.9	62.3	41.0	66.9	83.2	41.8	0.7
1986	Delete ban on divorce	60.8	67.7	51.0	36.5	58.5	20.8	0.6
1987	Ratify Single European Act	44.1	53.8	34.3	69.9	81.8	48.2	0.5

complete understanding of the Irish political situation must consider the role that Ireland plays in the world economy and, more particularly, in Europe. As stated earlier, Ireland emerged into independence as a fully fledged peripheral economy, tied unremittingly to British economic fortunes. Immediately upon independence, a debate began between the free traders (mostly important members of the Fine Gael party) and the autarchic nationalists, a group predominantly composed of members of the other two parties. While areas of large farms (east and midlands) were generally in favour of free trade and access to the British market, the west, south-west and border regions were in favour of stopping payments to Britain for land acquired from landlords in the late 19th century. On coming to power, Fianna Fail stopped the land annuities and presided over the protectionalist economy during the Depression and World War II years. (Ireland remained neutral during the war despite some pressure from the conservative wing of Fine Gael to join the UK and its allies.) The main economic policy was autonomous growth through import substitution. Predictably, this policy failed. (Lipietz, 1982, recounts why import substitution will not work for peripheral economies.) While Irish capital left the state in search of higher profits elsewhere, some small affiliates of British companies survived serving the tiny Irish market behind the tariff walls. Low-level industrialization occurred in the 1930s and 1940s in goods such as textiles, shoes, food products, car assembly and farm machinery. Wages remained low, agricultural products in unprocessed form constituted the main export earnings and emigration continued, peaking in the mid 1950s. Over two-thirds of Irish imports and exports were with Britain.

In 1958, the FF government formally changed its industrial policy, a shift in the politics of power in Taylor's (1985) terms. The economy was opened to foreign investment and competition as the state played an active role by subsidizing foreign industries through grants, tax breaks, worker training, and high expenditure on human capital, especially with major expansion of the educational sector. A remarkable transformation took place within two decades, hastened by Ireland's entry into the EEC in 1973. Perrons (1985) accurately described it as the success of a regime of intensive accumulation or Fordism. For multinational firms, Ireland was particularly attractive as an internal periphery, with free access to the large European market and with a cheap and flexible labour force. The import penetration of foreign products on a massive scale drove native industries into bankruptcy, especially in the traditional clothing, car assembly and food sectors. Comparing 1973 and 1978, the native proportion of sales in the clothing sector went from 73 to 49 per cent, of footwear from 63 to 34 per cent and of hosiery from 69 to 39 per cent (NESC, 1982). Currently, over half of all workers in Irish industry are employed by foreign firms. The rapid change-over of companies in the past decade and a paucity of links between foreign industry and the native economy (Ireland is essentially an export platform for these firms) have led to a questioning of the industrial development policy that subsidizes foreign capitalists at the expense of native labour (*The Economist*, 1988). In the 1980s unemployment and emigration have skyrocketed to the mid 1950s levels and most general elections have been fought on the economic issue. Recently, after the Fianna Fail success in the February 1987 election and with the support of the PD and FG parties, the government has adopted an aggressive line on encouraging native enterprise while pursuing a generally 'Thatcherite' economic policy.

Along with the economic changes came social change on a massive scale. Ireland in 1955 had more in common with the Ireland of 1905 than with the Ireland of 1970. As the centre of government Dublin has grown to a third of the country's population as the west and border regions continue to lose population. FF managed to penetrate the Dublin electorate early and has always contested the inner-city and council estate (public housing) votes with the socialist parties while splitting the middle-class vote with Fine Gael. Population shifts have not resulted in dramatic electoral changes. The most noticeable development was the revival of FG in the 1970s on the strength of the new bourgeois vote, especially in suburban Dublin.

The debate over entry to the EEC in the early 1970s clarified the economic split in Ireland, though surprisingly it said far less about the strength of traditional nationalist feelings. The case for EEC entry was put by the two main parties, FF and FG, the main farming and employer organizations and all four national newspapers. Key elements of their case were that Ireland could not remain an isolated economy and still be attractive to foreign investment, and since Britain was joining, Ireland also had to join. Irish capitalists saw it as a chance to broaden their market options and the farming community wished to take advantage of the guaranteed price support schemes of the EEC. The Labour Party and some trade unions opposed entry, as did groups of small farmers and a number of nationalist cultural organizations (Manning, 1978: 207). The turnout in the 1972 referendum was high (71 per cent) and the 'yes' vote, at 83 per cent was decisively in favour of entry. As in Britain (Kirby and Taylor, 1976), Norway (Valen, 1973; Hellevik and Gleditsch, 1973), Denmark (Petersen and Alklit, 1973), and Spain (Preston, 1979), national referenda on major constitutional questions have produced patterns of voting based on the core–periphery, rural–urban, and subordinate– dominant culture cleavages identified by Lipset and Rokkan. In Ireland the arguments of the core (stereotyped as middle-class south suburban Dublin voters) carried the day convincingly.

Surprisingly, since its future was at the core of the split in the nationalist movement in 1922, Northern Ireland has never occupied a prominent place in Irish electoral life. All major parties are committed to a peaceful reintegration of the province into the republic, and the removal of Article 44 (special position of the Catholic church) in 1972 was designed to make this integration easier. The Progressive Democrats have drafted a new constitution that removes Ireland's claim to sovereignty over Northern Ireland, and the issue may eventually be put to a national referendum if this party continues to prosper. However, Ireland has a history of rapid rises and declines of small parties, and this fact should give anyone who sees Irish politics as realigning pause to reconsider.

The political geography of Irish referenda, 1937–87

In some respects, referenda results seem to reverse normal expectations in Irish politics. The most hotly debated issues, such as the anti-abortion amendment, have produced relatively low turnouts and the general election trend of high turnouts in the west and other rural areas and low turnouts in Dublin are reversed in referenda. A third feature that is somewhat surprising is the high level of spoilt votes in 1937 and 1972. In 1937 the referendum was held at the same time as the general election and clearly many general election voters opted out of the referendum exercise. The highest turnouts have occurred when the parties were actively campaigning for a mandate, as in the 1937 plebiscite, the 1972 EEC referendum, and the 1986 removal of the ban on divorce. Only the EEC referendum has shown a turnout close to general election rates and that was an issue that offered a clear break with half a century of a dominant nationalist–sovereign and isolationist tradition. It is noticeable that constituencies with the highest turnouts are those that perceive their gains or losses to be the greatest from the proposed changes. The commercial farming areas (centre and south of the country) have led in turnout in EEC-related referenda while Dublin and Cork have led in turnout on the moral–religious issues. The poorest farming areas (on the geographical fringes of the country) have consistently had the lowest referendum turnouts, a noted contrast to the intensity and longevity of party loyalties in these regions. Murphy (1985) suggested that campaigning by local party officials is the main reason for the turnout variation as well as the relative levels of support for each issue. In the absence of national party positions, these local activists have an even greater ability to modify the expected vote pattern. In the brokeraged Irish electoral system, local politicians have much more influence than is usual in European democracies.

All referendum margins, with the exception of the 1937 plebiscite, have been substantial. Combining the yes vote with the turnout and the spoilt vote ratio for 1937 shows that only 38.6 per cent of the electorate supported the de Valera constitution, hardly a ringing endorsement (Sinnott, 1988). While there is some consistency in the voting patterns, the correlation coefficients do not allow us to predict local choices with any certainty. For example, the coefficient between support for the anti-abortion amendment and deleting the ban on divorce is only -0.36. Clearly some areas could support an anti-abortion amendment but also allow divorce. It is relatively difficult to identify an electoral mosaic that is consistent from referendum to referendum. Unlike the heterogeneous cultural and class-segregated societies of other Western nations, the Irish electorate is relatively homogeneous, socially unsegregated and retains a politics that has a weak social basis, so that there has been no increase in the standard deviation of either party support or referendum outcomes over time. The evidence in Table 5.2 shows that the most controversial moral–religious questions produced the largest regional deviations. The higher the support for the issue, the lower is the regional deviation. The nationalization hypothesis, that divergent local interests are gradually merged into a national consensus over time, is not supported in Ireland. From the first time it contested the Dail elections in 1927, the largest party, Fianna Fail, had a fairly uniform national spread and despite being a peripheral party, easily penetrated the core area (Dublin) from the 1930s (Garvin, 1978).

It is well established that the Labour Party shows consistently the highest ecological correlates with social and economic variables (Gallagher, 1985). By contrast, the two main parties, FF and FG, rarely show coefficients of determination with these explanatory variables greater than 0.30. This difference also appears in the correlation of party support and referenda choices (Table 5.2). In each case, the voting return for the general election closest to the date of the referendum was chosen. On the referenda questions, the Labour Party has been both consistent and vocal. It opposed both entry to the EEC and signing the Single European Act, arguing that the latter compromised Irish neutrality. It supported the divorce amendment and opposed adding the anti-abortion article to the constitution. It opposed de Valera's constitution in 1937 and supported deleting Article 44. Only on the latter issue does the ecological coefficient run counter to the expected sign. Labour's support base is working-class, both rural and urban, and these class issues are reflected in a rejection of the European link and a loosening of the traditional Catholic orthodoxy in the social/religious arena. In Galway, Limerick and parts of working-class Dublin and Cork, individual Labour and Workers Party Dail members were especially active in getting out the vote (Murphy, 1985). Their relative success is

Table 5.2: Statistics of referendum results and party correlates

Year	Issue	No. of constituencies	Mean yes %	Standard deviation	Regression with % FF		Correlation with % Labour[+]
					r	b	
1937	Plebiscite	34	50.9	7.1	0.81	0.607	-0.15
1972	Accession to EEC	42	82.7	4.4	0.26	0.154	-0.80
1972	Delete Article 44	42	79.8	3.9	-0.00	-0.002	-0.19
1983	Anti-abortion clause	41	67.8	12.3	0.55	1.205[*]	-0.55
1986	Ban on divorce	41	35.1	10.3	0.26	-6.341[*]	-.42
1987	Ratify Single European Act	41	69.3	8.2	0.12	3.249[*]	-0.34

[*] Regression coefficient significant at the 0.05 level

[+] Total of Labour and Worker Party votes in 1983, 1986, and 1987

illustrated by large regression residuals on Figs. 5.1–4, as the vote totals were unexpectedly higher or lower than the FF support in the area would suggest.

The highest ecological correlations appear for the referenda on which FF and FG took different sides (Table 5.2), on the 1937 plebiscite, the anti-abortion amendment, on which the FG leadership took a neutral position, and the divorce article (FG supported its deletion). The stronger the national support for the majority position, the weaker the FF correlation coefficient, representing the 'catch-all' nature of the two big parties and their widespread national patterns of support. Over time, it appears that the constitution has been viewed less as an FF party document and its principles and new amendments are now supported by both big parties and by the majority of Irish citizens. For the non-significant coefficients in Table 5.2 (the 1972 votes), significant correlations appear when the FF and FG votes are added together. The r values are 0.73 and 0.30 for the EEC and Article 44 votes respectively. Only the small Labour Party has consistently been on the losing side.

The residual maps are quite consistent from referendum to referendum. The area of the country that consistently adopted a core position (pro-Europe and liberal on moral–religious issues) is south suburban Dublin. The commercial farming areas of the centre, south and mid west strongly supported both EEC entry and the signing of the Single European Act but adopted traditional positions on the social–religious questions. The periphery of the country was traditional on both issues but was unsuccessful in guarding the idealist (de Valera) position of Irish sovereignty. Finally, the working-class areas of Dublin, Limerick, and Cork opposed links with Europe while supporting liberalization of the constitution on social questions.

The map of residuals from the regression of voting in the 1937 constitutional referendum (Fig. 5.1a) is particularly interesting since voters had to choose parties and vote to support or reject the constitution at the same time, though not on the same ballot. FF received less support nationally than did the de Valera constitution. The strongly nationalist counties on the border with Northern Ireland and three rural Cork and Kerry constituencies (traditional republican strongholds in the south-west) gave less support to the constitution than would be expected on the basis of their FF loyalties. For some voters in these areas, the constitution did not go far enough and the ratio of spoilt votes was also highest in these areas (O'Leary, 1979). It is worth remembering that Fianna Fail emerged from a split in the anti-treaty forces in the 1920s and that the Sinn Fein rump of that opposition was still strong in the border and southern areas. Their idealist form of nationalism believed that any constitution was premature until Northern Ireland was reunited with the South. In the two largest cities, Dublin and Cork, and in two

commercial farming areas in the centre of the country, the support for the constitution was surprisingly high as some voters rejected FF but accepted the notion of a constitution to seal the independence acquired 15 years earlier.

The two referenda (on different days) in 1972 allow a comparison of votes on both the social and sovereignty/economic questions. The broad patterns of the residuals are similar. On the EEC entry map (Fig. 5.1b), the Donegal North-east constituency appears as a strong residual because its leading politician had recently left the FF party and continued his career as an independent nationalist, thereby reducing the party vote in the area. Galway West and republican areas of the south-west as well as the working- class areas of Dublin did not favour EEC entry as much as would have been expected from their FF vote while commercial farming areas of the midlands and west strongly supported entry. On the religious question of deleting Article 44, on which there was unanimous party agreement and on which the Catholic church generally remained neutral, the residual pattern reflects the operation of local bishops. The Bishops of Cork and Limerick intervened, arguing against the removal of the article in their pastoral publications, while the Bishop of Galway was supportive of the removal (Murphy, 1985). This expression of 'pulpit politics' is clearly shown by the residuals in Fig. 5.2a. As expected, the strong FF middle-class areas of north Dublin also supported the article's removal.

On Figs. 5.2b, 5.3a and 5.3b the split between middle and working-class areas of Dublin on the economic issue and their agreement on social issues is clearly shown by the residuals. (Although the Single European Act issue was mostly a political question of sovereignty with implications for Irish neutrality, the debate was mostly about the economic costs and benefits of EEC membership.) On the moral–religious questions of divorce and anti-abortion, the suburban Dublin districts with Kildare (an increasingly suburban area immediately to the west of Dublin) appear as strong residuals (higher yes vote on the divorce question and stronger no vote on the anti-abortion clause than the FF proportion would suggest) but not the inner-city districts. Working-class areas of Dublin are clearly demarcated by age of households with older residents in the inner city and young families in large housing estates on the periphery. This age difference is evidenced on the maps. Elsewhere, Galway West and Clare (in the west) stand out as regional anomalies because of the effects of a well known Labour activist and Dail member in the area (Murphy, 1985). The south-west and the midlands appear as the most traditional regions on these social questions. Finally, the strongest positive residuals for the referendum on the Single European Act in 1987 are in the wealthiest Dublin constituencies and small farming areas while the

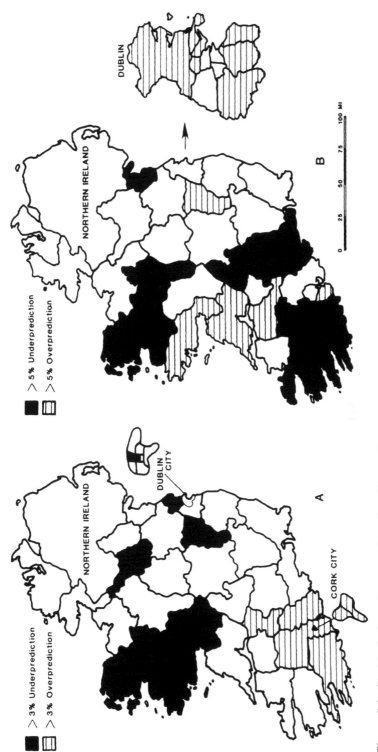

Figure 5.2: Residuals from the regression of the Fianna Fail vote on (a) Article 44, 1972, (b) inserting the anti-abortion clause in the constitution, 1983

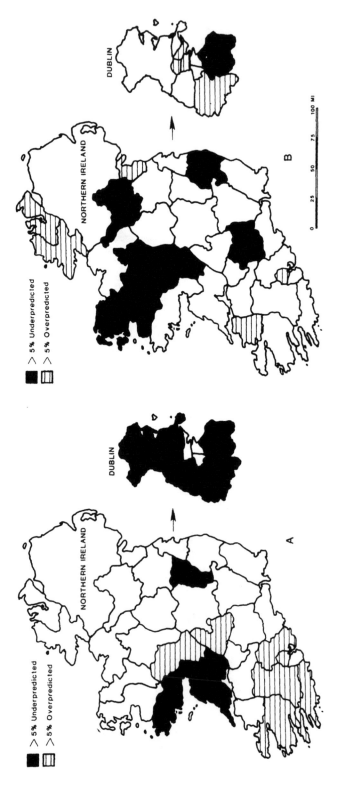

Figure 5.3: Residuals from the regression of the Fianna Fail vote on (a) the divorce referendum vote, 1986, (b) the vote on signing the Single European Act, 1987.

negative residuals are in working-class areas of Dublin and Cork as well as the nationalist strongholds of Donegal, Louth, and Kerry.

The results of this ecological analysis generally support previous studies of Irish general and special elections (Garvin and Parker, 1974; Walsh, 1984; Girvin, 1987; Murphy, 1985). What was not clear from these studies but became evident here is the appearance of two cleavages in Irish politics, a core–periphery cleavage and a capitalist–worker cleavage on both the national sovereignty and moral–religious questions. The political geography of these cleavages is well defined for Dublin and most of the country with some constituencies occupying middle positions both politically and geographically. The Fianna Fail party has been on the winning side in every referendum and thereby solidifies its claim as the national party. With over 40 per cent of the national vote and broad support in all sections of the electorate, the position of this party is critical in determining the odds of success of referenda.

Spatial associates and local contexts in Irish referenda

One of the major debates in electoral geography concerns the role of local context in shaping electoral choices. Geographers, led by Johnston (1985, 1987) and supported by the work of Savage (1987), claim that electoral studies that do not place the voting decision in context (usually meaning neighbourhood but sometimes constituency) are ignoring an important explanatory element. Because of the socializing effect of residence in an area, the expected electoral choices based on class status are skewed. McAllister (1987) and other political scientists argue that the contextual effect is nothing more than the failure to incorporate all necessary explanatory variables into the analysis and when this is done the so-called 'geographic effect' disappears. In Western democracies the voting pattern is expected to exhibit regional concentrations, especially if the centre–periphery cleavage identified by Lipset and Rokkan occurs. We therefore have two possible geographic patterns: a random pattern reflecting the specific combination of the characteristics of the district population and other local elements, and second, a clustered or regional pattern which reflects cleavages based on core–periphery, subordinate–dominant culture, or rural–urban differences. Recent advances in spatial modelling allow the measurement of clustering and spatial association in order to determine whether regional or local patterns are operating. Moran's I is an index of spatial autocorrelation that measures whether a geographic pattern differs significantly from a random distribution. It is analogous to temporal autocorrelation and is calculated from the ratio of covariance of the spatial units, using the weighted sums of the distances between district

Table 5.3: Spatial distribution and spatial association of referendum and party votes

Year	Issue	Moran's I		Tjostheim index	Association % Yes and FF %	
		% Yes	% FF		Z score	Probability
1937	Plebiscite	-0.029	0.021	0.020	0.544	0.29
1972	Accession to EEC	0.185*	0.127*	0.061	2.029	0.03*
1972	Delete Article 44	0.081*	0.127*	0.010	-0.316	0.61
1983	Anti-abortion clause	0.283*	0.205*	0.064	1.627	0.06
1986	Ban on divorce	0.380*	0.110*	0.056	1.603	0.06
1987	Ratify Single European Act	0.109*	0.110*	0.062	1.813	0.04*

* Significant at the 0.05 level.

centroids, and the variance of the variable of concern, here either the Fianna Fail or the referendum vote percentages. (See O'Loughlin, 1982 and Kirby and Ward, 1987, for the method and some political geography case studies.)

The Moran's I values for the referendum results and Fianna Fail are shown in Table 5.3. Except for the 1937 figures, all the election vote patterns show significant spatial clustering (difference from an expected random pattern). The more controversial the issue, the stronger is the clustering effect as social and economic cleavages in Irish society rise to the surface. By far the largest Moran's I values are for the anti-abortion and divorce referenda, as Dublin and the rest of the country diverged on these moral–religious issues. This spatial clustering is far more than the Fianna Fail vote as is seen in the Moran's I values for the comparable general election results in Table 5.3. The moral–religious questions have elicited a regional sharpening of the traditional–modern cleavage in a way that is not apparent in votes on entry to Europe. While the major parties agreed on the European question, they split on moral–religious issues, showing that the divisions of the 1920s into idealist and compromising nationalist forms have not dissipated.

Recent work on geographic association has identified a method of determining how much of the ecological correlation between two variables is due to the values of the variables at the places and their paired comparison and how much is due to the spatial arrangement of the places with respect to each other. In other words, for this study, what is the probability that one could obtain a correlation coefficient between Fianna Fail and referendum votes as large as that given by the existing spatial arrangement of constituencies, if the set of constituencies were randomly permutated through all possible locations? The method was first developed to detect space–time patterns of cancer rates but it is equally applicable to any geographical situation. Formally, for a set of N objects, the method involves the definition of a pair of matrices, C and Q, each of order N x N, the number of constituencies. In this study, one similarity matrix is a distance matrix between constituency centroids; the other is the difference in constituency ranks on the Fianna Fail and referendum ('Yes') percentages. To test the relationship between the similarity matrices, we compute the Tjostheim statistic which is the sum of the cross-products, and a method for finding the associated significance level (the probability of a value as large as this occurring by chance) is now available (Costanzo, 1983; Hubert *et al.*, 1985).

There are two types of association present in the voting data for Irish elections. First, there is a point-to-point correspondence, measured by Pearson's correlation coefficient between party and referendum votes as in Table 5.2. The second is measured by the Tjostheim index and summarizes spatial association and it is based on how the components defining these point-to-point association measures are themselves spatially arranged.

Only two elections have a Tjostheim index significant at the 0.05 level, the 1972 vote on accession to the EEC and the 1987 vote on the signing of the Single European Act, but the spatial association between the FF and referendum votes is significant in two other instances, the anti-abortion vote in 1983 and the divorce ban removal vote in 1986. For a significant Tjostheim index, a strong point-to-point correlation coefficient must be present and the spatial clustering of both votes would have to be spatially clustered in a similar way, as happened in both 1972 and 1987. In both European votes, the strong Labour areas of Dublin differed from the national majority opinion. On the two

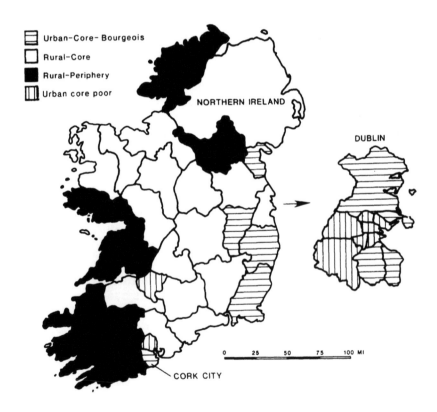

Figure 5.4: The political geography of the Lipset–Rokkan cleavages in the Republic of Ireland

moral–religious issues all Dublin constituencies, except the central city, had vote patterns dissimilar to the rest of the country by opposing the insertion of an anti-abortion clause in the constitution and supporting the removal of the divorce ban. The other two referendum votes were essentially the result of local factors, such as the party composition of the district, the level of campaigning and the effort of local activists, and the relative level of adherence to party platforms.

Lipset–Rokkan cleavages appeared in four of the six Irish referenda, resulting in regionally consistent patterns of voting. The first is essentially the worker–capitalist cleavage appearing in Irish referenda as sovereignty–economic votes and the other, a centre–periphery cleavage, concerns the maintenance of the traditions enshrined in the constitution despite a half-century of social change. The Lipset–Rokkan cleavage formation model offers a useful starting point in examining the results of Irish referenda. The particularities of the Irish party system, however, preclude a general acceptance of the model in every electoral contest.

The geography of Lipset–Rokkan cleavages in Ireland

From the results of the referenda and the correlations with party voting, as well as turnout data, it is possible to classify Irish constituencies according to the regional geography of Lipset and Rokkan's cleavages. On the core–periphery cleavage, Dublin is clearly distinguished from the rest of the country. On the moral-religious front, Dublin is basically a politico-geographic unit, with both bourgeois and working-class districts sharing a liberal position; only small pockets in the other cities share Dublin's position. Therefore, two Lipset–Rokkan cleavages, urban–rural and core–periphery, are clearly evident in results of the Irish referenda over the past decade and a half. On the economic front, brought into clear relief by the votes on EEC entry and signing the Single European Act, the Dublin core splits and the national cleavage becomes a capitalist–worker division. The Dublin middle-class areas are joined by the commercial farming interests and by the bourgeoisie of the other towns to form a clear majority over the Labour/Worker Party strongholds and the nationalist periphery. This functional axis, in Lipset and Rokkan's terms, is becoming more pronounced in the current climate of economic difficulties as two new parties, adopting unequivocal positions on the economic front, have prospered. The Workers Party advocates a socialist future and is rapidly gaining in the usual Labour strongholds while the Progressive Democrats are taking votes disproportionately from Fine Gael while sharing the same unrestrained capitalist principles.

The geography of these cleavages is shown in Fig. 5.4. The first set of 10 constituencies is the core/urban and bourgeois areas in suburban Dublin, in the nearby urbanized counties and in the southern part of Cork. They have been defeated on the moral–religious issues but are part of the winning coalition on the economic front with most of the rural periphery. It is this group of districts that increasingly sets the tone and direction of Irish politics and the three centre-right parties are currently vying for its loyalty. The second set comprises 14 constituencies of 'middle Ireland' from the north-west to the south-east. Primarily rural and small-town localities, Fianna Fail and Fine Gael's dominance has been consistent, with Fine Gael stronger in the centre and south of this region, based on their appeal to the large farming class and small-town businessmen. The turnout rates of these constituencies are the highest in both general elections and referenda, though they have been surpassed by the Dublin districts in the referenda of 1986 and 1987. On both the moral–religious and economic issues this group of counties is part of the winning coalitions, in the former with the traditional areas farther west and in the latter with the bourgeois areas of the urban core. Electoral allegiances are well defined in this area and the parties are remarkably consistent in their first-preference percentages from general election to general election.

A third group of constituencies is comprised of 9 peripheral poor districts on the fringes of the country. This is the archetypical nationalist stronghold and encompasses the Irish-speaking Gaeltacht areas. Fianna Fail generally dominates this area, but Fine Gael, as well as more idealist nationalist parties, have taken votes from Fianna Fail. On the Lipset–Rokkan cleavages, this group of counties has been more reluctant to support Irish economic integration into Europe and has been almost as conservative as the rural/small town areas to the east on the moral–religious questions. The border constituencies, in particular, respond to events in Northern Ireland as indicated by the intermittent history of successful parliamentary candidates from Sinn Fein.

The final group of constituencies, 8 in total, is the urban/core working-class areas. Politically, they are divided between Fianna Fail and Labour (and, increasingly, the Workers Party). In general elections the turnout is low in these districts, but in recent referenda these areas have been among the highest, due in no small measure to active campaigns against European unity and for divorce by local socialist activists. These districts have the most consistent history of failure in recent Irish referenda, being outvoted by the rural coalition on the moral–religious issue and by the urban middle-class/commercial farming coalition on the question of European entry.

Conclusions

This chapter has found considerable support for the usefulness of the Lipset–Rokkan thesis that political cleavages were formed in the 1920s in Europe and remained essentially frozen. In Ireland the mobilizing election was in 1918 and it occurred in the context of a peripheral revolt against the British metropole. The parameters of the new state were fixed in 1937 by a constitution that came under considerable pressure for revision when the government tried to change the country's role in Europe and in the world economy after 1958. Constitutional changes allowing the state to accede to the European communities have been easily passed on the grounds that Ireland would benefit economically from entry. On the more controversial moral–religious questions the traditional position has been upheld and even strengthened.

In Taylor's (1985) terms, changes in the politics of power concerning Ireland's role in the world economy have managed to generate a politics of support that is successful on economic issues but not on social/religious issues. As a result, a fifth cleavage, on the issue of the external relations of the state, should be added to the four cleavages identified by Lipset and Rokkan. The ideas of the core have not been able to defeat the bloc of voters which still adheres to the social and religious traditions of de Valera's constitution. The end result is that while Ireland is integrated economically with Europe, be it as an internal periphery, Ireland remains an isolated state on social issues when compared to the other member states of the European Communities. How long this distinction can persist remains to be seen. Cautiously, it may be observed that the recent successes of two new parties, the Progressive Democrats and the Workers Party, vastly different on the economic issues but both socially liberal, may be indicative of change in the future.

References

Carty, R. K. (1981) *Party and Parish Pump: Electoral Politics in Ireland*. Waterloo, Ont: Wilfrid Laurier Press.

Chubb, B. (1970) *The Government and Politics of Ireland*. Stanford, CA.: Stanford University Press.

Costanzo, C. M. (1983) 'Spatial inference in geography: modern approaches spell better times ahead', *Professional Geographer*, 35: 158–64.

Economist (1988) 'The poorest of the rich: a survey of Ireland'. 16 January: 1-26.

Farrell, B. (1970) 'Labour and the Irish political party system', *Economic and Social Review*, 1: 477–502.

Gallagher, M. (1976) *Electoral Support for Irish Political Parties*. London: Sage.

—— (1985) *Political Parties in the Republic of Ireland*. Dublin: Gill & Macmillan.

Garvin, T. (1978) 'The destiny of the soldiers: tradition and modernity in the politics of de Valera's Ireland', *Political Studies*, 26: 328–47.

—— (1981) *The Evolution of Irish Nationalist Politics*. Dublin: Gill & Macmillan.

—— and Parker, A.J. (1974) 'Party loyalty and Irish voters: the EEC referendum as a case study', *Economic and Social Review*, 4: 35–41.

Girvin, B. (1986) 'Social change and moral politics: the Irish constitutional referendum 1983', *Political Studies* 34: 61–81.

—— (1987) 'The divorce referendum in the Republic: June 1986', *Irish Political Studies* 2: 91–96.

Hellevik, O. and Gleiditsch, N.-P. (1973) 'The Common Market decision in Norway: a clash between direct and indirect democracy', *Scandinavian Political Studies* 8: 227–35.

Hubert, L. J., Golledge, R. G., Constanzo, C. M. and Gale, N. (1985) 'Tests of randomness: unidimensional and multidimensional', *Environment and Planning A* 17: 373–85.

Johnston, R.J. (1985) *The Geography of English Politics: the 1983 General Election*. London: Croom Helm.

—— (1987) 'The geography of the working class and the geography of the Labour vote in England 1983', *Political Geography Quarterly* 6: 7–16.

Kirby, A. M. and Taylor, P. J. (1976) 'A geographical analysis of the voting pattern in the EEC referendum, 5 June 1975', *Regional Studies* 10: 183–92.

Kirby, A. M., and Ward, M. D. (1987) 'The spatial analysis of peace and war', *Comparative Political Studies* 20; 292–313.

Laver, M. (1986) 'Ireland: politics with some social bases; an interpretation based on aggregate data', *Economic and Social Review* 17: 107–31.

—— (1987) 'Measuring patterns of party support in Ireland', *Economic and Social Review* 18: 95–100.

Lijphart, A. (1971) 'Class voting and religious voting in European democracies', *Acta Politica* 6: 158–71.

Lipietz, A. (1982) 'Towards global Fordism?', *New Left Review* 132: 33–47.

Lipset, S. M. and Rokkan, S. (1967) 'Cleavage structures, party systems and voter alignments: an introduction', in S. M. Lipset and S. Rokkan (eds.) *Party Systems and Voter Alignments*, New York: Free Press, 3–64.

McAllister, I. (1987) 'Social context, turnout and the vote: Australian and British comparisons', *Political Geography Quarterly* 6: 17–30.

Mair, P. (1987) *The Changing Irish Party System*. London: Frances Pinter.

Manning, M. (1978) 'Ireland', in D. Butler and A. Ranney (eds.) *Referendums: a Comparative Study of Practice and Theory*. Washington DC: American Enterprise Institute, 193–210.

Marsh, M. (1985) 'Ireland', in I. Crewe and D. Denver (eds.) *Electoral Change in Western Democracies*. London: Croom Helm, 267–86.

Morgan, A. (1988) *James Connolly: a Political Biography*. Manchester: Manchester University Press.

Murphy, M. (1985) 'A Spatial Analysis of Irish Referenda, 1959-84', unpublished B.A. (Hons.) dissertation, Department of Geography,

University College, Dublin.

NESC (National Economic and Social Council) (1982) *A Review of Industrial Policy: a Report prepared by the Telesis Consultancy Group*. Dublin: NESC.

O'Leary, C. (1979) *Irish Elections, 1918–1977*. Dublin: Gill & Macmillan.

O'Loughlin, J. (1981) 'The neighbourhood effect in urban voting surfaces: a cross-national analysis', in A.D. Burnett and P. J. Taylor (eds.) *Political Studies from Spatial Perspectives: Anglo-American Essays on Political Geography*. Chichester: Wiley, 357–88.

Parker, A. J. (1982) 'The "friends and neighbours" effect in the Galway West constituency', *Political Geography Quarterly* 1: 243–62.

—— (1986) 'An ecological analysis of voting patterns in Galway West, 1977', *Irish Geography* 17: 42–64.

—— (1987) 'Breaking the mould?: The Irish general election of February 1987', *Irish Geography* 20: 50-53.

Penniman, H. and Farrell, B. (eds.) (1987) *Ireland at the Polls; 1981, 1982 and 1987: a Study of four General Elections*. Washington, D.C.: American Enterprise Institute.

Perrons, D. (1985) 'Unequal integration in global Fordism: the case of Ireland', in A. J. Scott and M. Storper (eds.) *Production, Work, Territory: the Geographical Anatomy of Industrial Capitalism*. London: Allen & Unwin, 246–64.

Petersen, N. and Alklit, J. (1973) 'Denmark enters the European Communities', *Scandinavian Political Studies* 8: 198–213.

Preston, P. (1979) 'The Spanish constitutional referendum of 6 December 1978', *West European Politics* 2: 246–49.

Sacks, P. (1970) 'Bailiwicks, locality and religion: three elements in an Irish Dail constituency election', *Economic and Social Review* 1: 531–54.

Savage, M. (1987) 'Understanding political alignments in contemporary Britain: do localities matter?', *Political Geography Quarterly* 6: 53–76.

Sinnott, R. (1984) 'Interpretations of the Irish party system', *European Journal of Political Research* 12: 289–307.

—— (1988) 'The Referendum in Ireland: Constitutional Provisions and Political Practice', unpublished paper, Department of Politics, University College, Dublin.

Taylor, P. J. (1985) *Political Geography: World-Economy, Nation-State and Locality*. London: Longman.

—— and Johnston, R.J. (1979) *The Geography of Elections*. Harmondsworth: Penguin.

Valen, H. (1973) 'Norway: no to EEC', *Scandinavian Political Studies* 8: 214–26.

Walsh, B. (1984) 'The influence of turnout on the results of the referendum to amend the constitution to include a clause on the rights of the unborn', *Economic and Social Review* 15: 227–34.

Whyte, J. H. (1974) 'Ireland: politics without social bases', in R. Rose (ed.) *Electoral Behavior: a Comparative Handbook*. New York: Free Press, 619–51.

Chapter six

Volatile stability: New Zealand's 1987 general election

Rex Honey and J. Ross Barnett

New Zealand's 1987 election looks at first glance remarkably like the preceding election in 1984. The Labour Party emerged with the identical 19-seat advantage over the opposition National Party that it held after the 1984 election. In addition the stability of party control over constituencies was unusually great. In fact, since Labour and National emerged as the dominant parties in the 1930s, only the 1963 election had fewer seats move from one party to another.

Beyond this first glance, however, are substantial changes, possibly fundamental changes in the geography of New Zealand politics. Consider, for example, Alan McRobie's (1987) analysis before the election. Using the conventional political analysis of a uniform national swing, McRobie said the National Party needed a 4.7 per cent two-party swing to win the election and that Labour needed to prevent a swing of as little as 3.3 per cent to assure retention of power. In fact, National achieved a 4.5 per cent two-party swing, enough, according to McRobie's analysis, to deadlock parliament and possibly enough to catapult the challengers back into power. Despite the loss of votes, Labour remain solidly in control rather than losing power. Analysis of a uniform national swing is instructive only (if ever) when basic political cleavages are stable. This obviously was not the case.

What, then, was the 'case'? What, below the level of appearances, transpired in New Zealand's 1987 election in terms of political geography? The geography of elections, after all, is part of political geography; as such it is best understood in terms of the context of socio-political struggle for power (Agnew, 1987). Arguably, this election – indeed all elections – is best understood in terms of social theory placing the experience in its historical-geographical context, considering the production and reproduction of ruling coalitions and cross-class allegiances (Harvey, 1985). New Zealand has many attributes making it a good laboratory for using elections to gauge political change. It is a parliamentary democracy with three year terms, so the voters have frequent opportunities to express their preferences.

New Zealand's role in the world economy has changed markedly in recent decades, with concomitant changes in the spatial structure of the country's political economy. Finally, New Zealand is a small country, at least in terms of population, so that politics is relatively personal.

The 1987 vote in historical perspective

Immediate reaction to the election was focused upon Labour's historic achievement of winning re-election for the first time in more than 40 years – in fact since the first Labour Government won its fourth term in a row in 1946. However, perhaps even more historically ·significant changes took place – changes in the basic geography of New Zealand politics. The next few elections will show whether they are historic transformations or merely ephemeral idiosyncrasies of a turbulent time.

One of these changes, building on voting trends evident in 1984, is a realignment of party political allegiances. Though from the perspective of other Western democracies New Zealand's dominant political parties have appeared to be quite similar (Johnston and Honey, 1988), the parties have long appealed to different sections of the country's socio-economic spectrum (Barnett, 1977; James, *et al.*, 1987). In short, Labour's traditional base is working class, especially in urban areas (including provincial cities, not just the metropolitan areas of Auckland, Wellington and Christchurch). The National Party has traditionally won support from wealthier voters, as well as from those in suburbs and rural areas.

The traditional socio-economic cleavage was in fact confused in 1984, as a consequence of changes in the country's political economy at both inter- and intranational scales. (Arguably the cleavages were weakening earlier, but the 1981 election was distorted by an unusual issue – civil unrest over a South African rugby tour.) Before the 1984 election the National Party had governed for nine years as well as 21 of the previous 24 and 29 of the last 35. The economy was in a shambles, and National naturally took the brunt of the blame. Most of the affluent (and many in the middle), though, mistrusted Labour; in 1984 almost one-third of the voters turned to third parties. Thanks to policies perceived broadly as pro-business, in 1987 many of the affluent and intendedly affluent voted Labour while many of the poor and working-class did not. The traditional cleavage, at least for the time being, weakened. Meanwhile the traditionally less important rural–urban split became even more pronounced. National solidified its margins in the countryside but almost disappeared in the major cities. At least in 1987, the rural–urban dimension supplanted the socio-economic one as New Zealand's pre-eminent voting cleavage.

The other major change is the relationship between votes and seats in the competition between the two main parties. Labour, as is often the case for parties on the left, has traditionally had a difficult time turning voting strength into electoral strength. Labour, for example, obtained more votes than National in both the 1978 and 1981 elections only to see National control parliament and therefore form the government, the true goal in a parliamentary election. McRobie's expectation for 1987, in fact, was that the National Party would win the election unless Labour had a 3 per cent vote advantage nationally. Labour out-polled National by only 2.6 per cent yet its real victory, i.e. advantage in parliamentary seats, was substantial. Never before had Labour won an election when the popular vote between the two parties was close. This time Labour effectively targeted the electorate in a carefully designed spatial strategy.

Essentially, New Zealand emerged from the 1987 election with a new electoral geography, reflecting both a transformation of the nation's spatial economy and an enlightened campaign by the Labour Party. Let us examine these changes in the light of the concepts and theories of electoral geography.

The normal left-wing disadvantage

New Zealand shares with many other Western democracies – Australia, Britain, Canada, Ireland and the United States – a constituency based electoral system with representatives chosen in first-past-the-post polls. Except in Ireland and the United States, where the process of drawing electoral boundaries is expected to engender political advantage, the constituency structure is supposed to be neutral. A party winning a majority of votes is expected to win a majority of seats – and therefore to govern.

In Australia, Britain, Canada and New Zealand, politically neutral boundary commissions determine the boundaries of constituencies, typically following local authority boundaries as much as possible to create constituencies with roughly equal population and communities of interest. (Among these countries, the Australian state of Queensland is infamous for a bias assuring the dominance of rural areas despite an urban population majority.) In New Zealand the Representation Commission determines new constituencies after each census, i.e. every five years.

Scholars have long recognized that this seemingly neutral, objective and just procedure actually grants an electoral advantage to a party with a widely distributed base of support relative to a party with a highly concentrated base of support (Butler, 1952; Rydon, 1957; Brookes, 1960; Taylor and Gudgin, 1977). Concentration of one party's voters is common in Western democracies when that party has working-class or minority support as a significant part of its total electoral support. Under such circumstances the constituencies take the form of and have the consequences of opponent concentration gerrymanders, despite the absence of motivation for electoral advantage (Austin *et al.* 1987).

New Zealand's Labour Party has a history of class-based support. Emerging as a potent force during the Great Depression of the 1930s, Labour championed the cause of the underdog in New Zealand society. Labour's strongholds have been inner-city working-class areas and the provincial cities of the South Island – relatively downtrodden places where Labour has traditionally won overwhelming majorities, because of opponent concentration gerrymanders.

Efficiency of New Zealand votes

The relationship between a party's ability to win votes (relative to its major opposition) and its ability to win parliamentary seats shows Labour's traditional problem of wasting votes (Table 6.1). Before 1987, only when Labour had a substantial vote advantage over National was it able to match that with an equal or greater advantage in seats. Heretofore, Labour succeeded at the polls only when achieving a substantial victory in votes, as in 1972 and 1984. National, by way of contrast, could turn a small advantage into a substantial victory. National has even been able to lose the popular vote, albeit very narrowly, yet win control of parliament; in 1978 that control was extraordinary, given Labour's actual edge in votes.

Table 6.1: Winning votes and winning seats in New Zealand elections 1954–87

Elections	Labour Victory		National Victory	
	Voting Margin %	MP margin %	Voting Margin %	MP Margin %
1954			0.15	12.50
1957	4.10	2.50		
1960			4.17	15.00
1963			3.33	12.50
1966			2.20	11.25
1969			1.04	6.90
1972	6.87	26.44		
1975			8.02	26.44
1978	0.57			11.95
1981	0.23			4.35
1984	7.09	20.00		
1987	2.60	19.58		

Note: Labour and National are compared against each other for each election in terms of the differences in their percentages of the total vote and their percentages of all parliamentary seats.

Typically, Labour has won the inner city electorates with overwhelming majorities, only to lose close elections in many constituencies elsewhere. Usually most marginals have been in National control most of the time. Prior to 1987 only when Labour succeeded in pushing its total vote to 6 or 7 per cent above National's did those marginals move to Labour's side of the aisle.

In historical terms, the 1987 election is a clear anomaly – a relatively narrow Labour advantage in votes but an overwhelming Labour victory in Parliament. This is the closest popular vote ever in an election Labour has won, yet Labour's parliamentary majority is the biggest it has had since the Great Depression. (In 1984 Labour had a total majority of 17 because the Social Credit Political League won two seats.) How did it happen? And what will happen next? To answer the first of these questions we must consider electoral theory, the issues of the 1987 campaign, and the spatial strategies of the campaign, as well as class based political theory. To answer the second we must consider these factors – and wait for elections in 1990 and beyond.

Electoral issues, theory, and the 1987 election

In 1984 Labour formed a new administration amidst crisis. The economy was a shambles; the New Zealand dollar was suffering in international markets, and a sharp break with the past was needed. When the defeated Prime Minister, Sir Robert Muldoon, begrudgingly yielded power, the fourth Labour government embarked on an economic programme differing from the interventionist policies of the National government as well as the socialist tradition of the Labour Party. The new economic policies, fashioned by the Finance Minister Roger Douglas, were monetarist following Friedman rather than socialist or even Keynesian (Boston and Holland, 1987; Douglas, 1987).

For decades New Zealand practised a self-proclaimed benevolent capitalism, usually under a National Party administration. New Zealand was an experimenter and innovator in such socially progressive policies as social security, public health, women's suffrage, and the protection of workers. Over the decades succeeding central governments intervened directly in the economy by taking over ailing enterprises (as in forestry and mining), investing in areas deemed to be of national interest (as in oil refining), and providing protection for domestic producers (particularly farmers but also many kinds of industry) (Sinclair, 1969; Oliver, 1981). Those policies were more the result of various coalitions of capital and labour than a grand design, more the product of political action through a particular social structure than an orchestrated plan. Kiwis regarded themselves as inventive, relatively well-off, and free.

They regarded their government's policies as simply assuring the just operation of a sound economy. Everyone had enough; those who would work could have plenty, but no one got really rich.

Events overcame the perception. In the generation after World War II New Zealand dropped from near the top of the list of OECD countries in terms of *per capita* GDP to near the bottom. The loss of a major market when the UK entered the EEC, the great increase in the price of petroleum, and New Zealand's distance from the world's major economies helped to shatter the Kiwis' self-satisfaction. By 1984 the New Zealand electorate was ready for change.

Rogernomics, as the new economic policies came to be known, transformed New Zealand's spatial economy and consequently its spatial politics. In Britain and the United States regimes regarded as the most right-wing in decades, if not this century, introduced monetarist policies. The Labour Party introduced them to New Zealand. Rogernomics transformed the spatial economy by replacing public control of the economy with an unfettered, unbuffered dependence on market forces. Support for agriculture, forestry, and mining evaporated, exposing entire sectors to competition (international competition, often subsidized in other countries) after generations of protection. Public policies engendering balance among the regions gave way to market-based economic concentration in places best able to compete under the new circumstances. Particularly successful have been Auckland, the fruit and wine producing areas, and regions with strong tourist amenities.

The literature on electoral theory concentrates on the efforts of political parties to attract and maintain a base of support. To assure themselves consistent and committed (to the point of working in campaigns) support, parties need the loyalty of certain sections of society. Opposing parties typically depend on different groups for their base support. The differences in these groups are the cleavages demarcating the structure of politics in a society (Lipset and Rokkan, 1967; Taylor and Johnston, 1979).

Typically, in constituency based electoral systems such as New Zealand's two parties dominate (Rae, 1971; O'Loughlin, 1980). When a new party breaks the dominance, one of the old ones either recedes, as Britain's Liberal Party did, or the old ones merge, as happened in New Zealand when the Liberal and Reform Parties merged to become the National Party following Labour's rise to power in the 1930s. The parties will battle, trying to gain enough middle ground to win elections without deserting their bases so much that they lose their political souls (Downs, 1957). Over time the nature of political issues changes, with consequent realignment of political cleavages. At the turn of this century the Republican Party was the more liberal of the United States' dominant parties, for example; Franklin Roosevelt's New Deal marked a major realignment of US politics (Archer and Taylor, 1981).

Rogernomics marked a major break with the past, not only giving Labour appeal to the economic elite alienated by National in 1984, but also straining Labour's ties with its working class power base. In terms of electoral theory, once in power Labour tried to broaden its base, moving away from its traditional core to gain new supporters among the economic movers and shakers. Labour usurped National's traditional turf by adopting policies attractive to Bob Jones's New Zealand Party supporters. (New in 1984, the New Zealand Party gained 12.3 per cent of the vote, though no parliamentary seats. Jones, a charismatic land developer, promised a more open economy and won the New Zealand yuppie vote. His supporters had mainly voted for National in earlier elections.) Labour exposed itself to future problems by trying to appeal to such a wide spectrum of voters that it risked gaining temporary voting support at the possible cost of diminishing among party workers the fervour so necessary to mount campaigns. In other words, Labour's adoption of monetarist economic policies could achieve ballot box success in the short run by attracting new voters with no real commitment to the party, only to see the policies reduce the party's cadre of workers sufficiently committed to toil for the party and be loyal to it.

Labour's strategy is a clear example of Harvey's changing coalitions of labour and capital for pro-capitalist purposes. Labour's strategy echoes Savage's (1987) analysis of British elections in that it affected the material interests of people occupying different positions in the country's social structure. Effectively, Labour wrote off the areas suffering from the new economic order: farming constituencies, as well as those based on forestry and mining. It was willing to risk this, in accordance with this view, because of potential support from the economic elite of the upper income suburbs. Labour leaders, of course, argued that any pain suffered would be temporary – or that it could be ameliorated by moving to places with greater economic viability.

Spatial strategy of the 1987 election

As soon as the Labour government had got its breath back after dealing with the economic emergencies facing it when it took office in 1984, the party began planning the 1987 campaign. Though the official campaign covered the three weeks running up to the 15 August election, that period really marked only the last stages of the campaign. Labour ran a very sophisticated programme involving the targeting of marginal seats, careful and continuous public opinion polls at constituency level, and highly localized advertising campaigns.

Having been out of office so long, Labour had a fresh team in the Cabinet. David Lange, the new Prime Minister, had achieved victory in his first election as party leader. Most Ministers had never been part of a Cabinet before. Labour projected a bold, innovative image. The Labour government sought to project an image of competence and caring – an administration competent to govern the country and manage the economy, yet compassionate in that it needed to manage well in order to be able to do the progressive things necessary to provide for everyone's basic needs.

The National Party, meanwhile, was in disarray. National's 1984 share of the vote was its lowest ever, less than 36 per cent. The party image had been battered. What had been the country's natural party of government came to be regarded as incompetent and without direction. Rifts within the parliamentary party consumed considerable energy. Muldoon, though discredited to the point of being a liability, retained sufficient strength to bring down his successor as party leader, as well as make life miserable for the man who led the National campaign in 1987, the new opposition leader, Jim Bolger.

Labour recognized that a New Zealand election is not a single election but rather separate elections in each constituency. Party leaders were not as interested in the total vote or national opinion as in support in each electorate. Polling on a constituency basis, Labour identified the constituencies expected to be closest, i.e. the key marginals. These were mainly suburban or in areas dominated by a small to medium sized city. Birkenhead and Glenfield are examples of the former, New Plymouth and the Hamilton seats examples of the latter. Given Labour's sizeable parliamentary majority going into the election (15 seats overall and 17 over National after a by-election loss in Timaru), Labour concentrated on holding the seats it already held as well as targeting a few vulnerable National seats. Labour leaders also recognized the difficulty New Zealand governments have had in retaining their majority, even when winning re-election. (Only once in the ten elections prior to 1987 had a party in power been able to retain its parliamentary majority.)

Well before the official campaign, Labour began a multi-faceted effort to win the marginals. Frequent polling gauged voter preferences on a range of issues. This helped Labour tailor the national campaign (and government policy) to these issues as well as helping the party prepare advertising blitzes specifically designed for the key constituencies. Party leadership also paid careful attention to the selection of particularly strong candidates in these areas. Furthermore, Cabinet Ministers visited the marginals on a regular basis, especially to highlight policies advantageous to an electorate and policies meeting favour, according to the polls, with the voters.

Economic problems almost scuttled Labour's campaign in late 1986 and early 1987. National actually moved ahead in one public opinion poll (James *et al.* 1987). Labour even had a financial advantage at one stage because the business community preferred a Labour victory.

Labour's left, appalled by the ideological shift of the party, squawked but had little choice. Some members bolted, running alternative candidates, but most at the lower socio-economic end saw little alternative to voting Labour – if they voted. The party knew it would win its safe seats even without campaigning on issues popular in those areas, e.g. economic intervention to reduce unemployment or increased support for those on welfare. Rather, the government stressed the eventual benefits to all once the restructured economy had begun to grow.

Meanwhile National had problems. The legacy of economic chaos from the Muldoon government, the leadership squabbles, and the loss to Labour of much of National's traditional base put the party in jeopardy not only of losing but of suffering a monumental loss. National, though targeting marginals, opted for a hard-hitting national campaign taking the offensive at every opportunity. They pounced on Labour gaffes, particularly a document about a secret agenda, as well as Labour's seeming insensitivity to those suffering from the structural transformation of the economy. (The Prime Minister's evident diffidence battered Labour's image as the party that cared.)

Labour, by contrast, opted for a relatively quiet official campaign, being seen governing while the fruits of the strategy in place over the previous two years matured. Labour did not want to frighten away potential middle class and business elite voters with a rousing campaign stressing the party's socialist roots, striving to maintain support on the left through its nuclear freeze policy and measures to promote racial harmony. Labour ran a positive television campaign, showing a country succeeding and creating a secure place for itself in the world. Labour concentrated its print media attention on the marginals. The Labour Party Office designed a series of newspaper advertisements pointing at the issues of significance in each constituency, particularly the marginals. One version appeared to be a personal statement from the relevant candidate. The statement for Wairarapa, for example, stressed the development of tourism, that for West Auckland transport. Both included statements about hospitals and schools, differing only in the names of facilities.

The night before the election, the political scientist Keith Jackson told a national television audience that Jim Bolger of National had led a splendid campaign while David Lange of Labour had frittered away his lead. If the New Zealand elections were simply exercises in summing up votes, Jackson would have been right. Bolger turned a potential disaster into a close call. But New Zealand elections are in fact 97 separate elections. Lange and Labour chose to win a substantial advantage in these. They feared that a hard-hitting campaign could give Labour more votes but at the cost of fewer seats. In one sense each party achieved its primary objective. Labour won a substantial parliamentary victory, but National clawed its way back to respectability among the voters.

Results of the 1987 election

The 1987 election needs to be understood not only in terms of who won where (though that understanding is important, otherwise Labour's domination of parliament while attaining just a 2.6 per cent edge in votes would not make sense, particularly in the light of historical patterns); it also needs to be understood in terms of the size of victories in the individual constituencies and how the distribution of voting has changed.

Labour clearly penetrated National's support of the middle-class and business interests. This is clearly evident in Table 6.2, which illustrates shifts in Labour support at the electorate and census tract level in Christchurch. In both cities Labour's gains were greatest in the most affluent electorates or neighbourhoods and weakest in its traditional areas of support. In Auckland, Labour support grew most rapidly in the 'blue ribbon' seat of Remuera (up from 24.3 per cent in 1984 to 47.5 per cent in 1987). Labour came within 2 per cent of claiming what had been a solidly National constituency. In Christchurch the picture was much the same. Labour's greatest gains occurred in the city's three most affluent neighbourhoods (up 14.4 per cent in Fendalton, 15.7 per cent in Holmwood, and 14.4 per cent in Merivale); the Fendalton electorate, within which these three neighbourhoods are located, still remained in National's hands, though by a slashed majority. While these shifts are dramatic divergences from traditional patterns, Table 6.2 provides evidence that they did not obliterate entirely the traditional class cleavage.

Table 6.2: The socio-economic cleavage in Christchurch in the 1980s: correlations between percentage voting Labour and measures of social class, 1981–87

	Professional occupations (%)	*Earning above $30,000 (%)*	*Production workers*
1981	-0.77	-0.80	0.79
1984	-0.78	-0.82	0.79
1987	-0.68	-0.75	0.72
Change 1981–4	-0.19	-0.27	0.16
Change 1984–7	0.61	0.60	-0.58

Professional occupations are the professional-technical and administrative-managerial categories in the 1986 New Zealand census.

While Labour courted the middle ground of New Zealand politics and took away much of National's middle class base, the government suffered only modest defections from its traditional base. Significantly, those in lower socio-economic positions did not tend to register protest votes. Rather, they simply failed to vote (Johnston, 1988). Voting turnout declined nationally by 4.6 per cent with the biggest decreases in such safe inner city Labour seats as Manurewa, Panmure and Christchurch Central. (These seats are largely held by Labour Cabinet Ministers, prompting one election night commentator to cite a rejection of the Cabinet, failing to consider that these seats still produced overwhelming Labour majorities.) Some of Labour's traditional voters chose to abstain, but they could not get themselves to vote National.

The 1987 results further deepened the rural–urban split in New Zealand. The cities are increasingly red, the countryside increasingly blue, in terms of party colours (though both parties are increasingly blue in terms of ideology). By taking two seats from the Democrats (formerly Social Credit), National was able to win six of the 22 seats in metropolitan Auckland, compensating for the loss of two metropolitan seats to Labour. In Christchurch and Wellington, out of 18 seats National held only the aforementioned Fendalton, and that possibly because Labour had failed to target it and marshal a sufficiently intensive campaign. Even in the provincial cities, Labour forged serious domination, 14 seats to five; among these, National did better in the smaller places, such as Timaru and Whangarei, rather than in the largest ones, such as Hamilton and Dunedin.

In rural areas, though, the story is the opposite. National captured 19 of the 21 rural constituencies, Labour succeeding only in the Tasman and West Coast electorates, in each case with a significantly reduced majority. National also dominated the constituencies with mixes of urban and rural population, taking eight to Labour's five.

As telling as the rural–urban split, though, is the changing set of marginal seats and safest seats. In 1984 Labour won ten seats by more than 7,000 votes, National none. This amounts to a lot of wasted votes when viewed with the objective of winning parliamentary seats. In 1987 Labour had such big wins only in the Maori seats, and even in those their majorities fell. (Four of the 97 constituencies were reserved for Maoris in 1987. When registering to vote people identifying as less than 50 per cent Maori can opt to vote in a Maori or regular constituency.) The six inner city seats returning majorities above 7,000 in 1984 were still safely Labour but not by as much in 1987. The difference was lower voting totals rather than switches to alternatives, whether National, the Democrats or one of the fringe parties. Meanwhile, National's hold on the rural seats tightened. It won two constituencies by more than 7,000 votes in 1987, and its majority went up in almost all its rural seats.

Eleven seats were decided by fewer than 1,000 votes in 1984, with National winning six of these. Three years later sixteen seats had majorities of less than 1,000, the Wairarapa seat being decided by a single vote out of some 20,000 (presumably the Labour candidate's own!). The parties split these marginals. More seats are competitive. Even among relatively safe seats, Labour won a lot more by 1000 to 4,000 votes than by the 4,000 to 7,000 votes common in 1984.

Future voting patterns

Since 1946 National has usually been New Zealand's party of government, Labour the party in opposition. National Party leaders would, of course, like to think that the future will be a return to this norm. They base their hopes on the softness of Labour's support. By stretching outside Labour's normal base, National Party leaders told us, Labour has been able to attain electoral success, but it has mortgaged the party's future. This is because the new Labour voters have little real commitment to the party. They may vote for it; almost certainly they will not work for it. This move, as illustrated by the 1987 Labour Party conference, has alienated the party faithful. (Leftists in the party gained control of the party apparatus and proceeded to lecture the Labour government on party tradition and its socialist origins. The Prime Minister, seemingly amused rather than deterred by this turn of events, characterized the party as social democratic rather than socialist.) Labour is risking the loss of its dedicated work force. The National expectation is that Labour's economic miracle will not occur, leading to a loss of enthusiasm among leftist supporters alienated by monetarist policies. (Public opinion polls in 1988, showing a double digit National advantage, lend credence to this view.)

Labour leaders, not surprisingly, foresaw a different future when they discussed strategy with us. They are gambling that Labour's policies will work – or at least be seen to be working. The most optimistic of them see Labour leading a grand coalition of progressive thought, leading a country with a stout economy and social justice – a country where businesses have the freedom to innovate so the country can prosper in the world economy, yet where the fruits of prosperity are shared.

In terms of electoral theory, National has for the time being been outflanked. Labour has a grand coalition of left and right. Such coalitions seldom last. Labour is banking on an aura of competence to see it through. National is betting that it will not. National will try to re-establish the socio-economic cleavage because the rural–urban cleavage gives Labour too much of an advantage. Country parties fare poorly in a metropolitan world (except where protected by malapportionment as in Queensland). Alternatively, National could try to win voters on a new cleavage – perhaps an antipodean moral majority or a law and order fix. National's task, one way or another, is to break the Labour hold over the last two elections in urban New Zealand.

In terms of Harvey's theory, New Zealand has been experiencing change in its pro-capitalist coalitions. Based on the material interests of various groups, capital and labour in particular locations have joined together to advance their interests. These coalitions appear to be volatile, in part because of the relative advantages accruing to particular places but also because of the anticipated advantages of various segments of capital and labour. The 1987 election results mirrored those of 1984, but the underlying structure was substantially altered. The scene is set for substantial changes in results in 1990 because the 1987 structure is unstable and the earlier structure is history.

References

Agnew, J. A. (1987) *Place and Politics: the Geographical Mediation of State and Society*. London: Allen and Unwin.

Archer, J. C. and Taylor, P. J. (1981) *Section and Party*. Chichester: Wiley.

Austin, C. M., Honey, R., and Baxter, C. (1987) *Human Geography*. St. Paul, Minn.: West Publishing.

Barnett, J. R. (1977) 'Regional protest and political change', in R.J. Johnston (ed.) *People, Places and Votes: Essays on the Electoral Geography of Australia and New Zealand*. Armidale: University of New England.

Boston, J. and Holland, M. (eds.) (1987) *The Fourth Labour Government*. Auckland: Oxford University Press.

Brookes, R. H. (1960) 'The analysis of distorted representation in two party, single member elections', *Political Science* 12: 158–67.

Butler, D. E. (1952) *The British General Election of 1951*. London: Macmillan.

Douglas, R. (1987) *Toward Prosperity*. Wellington: Fourth Estate Books.

Downs, A. (1957) *An Economic Theory of Democracy*. New York: Harper and Row.

Harvey, D. (1985) *The Urbanization of Capital*. Baltimore: Johns Hopkins University Press.

James, C. *et al.* (1987) *The Election Book*. Wellington: Allen & Unwin.

Johnston, R. J. (1988) 'The beginnings of realignment? Ecological analyses of the 1984 and 1987 New Zealand general elections' *British Review of New Zealand Studies* 1: 27–49.

—— and Honey, R. D. (1988) 'The 1987 general election in New Zealand: the demise of electoral cleavages?' *Political Geography Quarterly* 7: 363–68.

Lipset, S. M. and Rokkan, S. E. (eds.) (1967) *Party Systems and Voter Alignments*. New York: Free Press.

McRobie, A. (1987) 'Decision '87: the pendulum', *The Listener*, 15–21 August.

Oliver, W. H. (1981) 'The awakening imagination', in W. H. Oliver (ed.), *The Oxford History of New Zealand*. Wellington: Oxford University Press, 430-61.

O'Loughlin, J. (1980) 'District size and party electoral strength: a comparison of sixteen democracies', *Environment and Planning A*: 247-67.

Rae, D. W. (1971) *The Political Consequences of Electoral Law*. New Haven: Yale University Press.

Rydon, J. (1957) 'The relation of votes to seats in elections for the Australian House of Representatives', *Political Science* 49-61.

Savage, M. (1987) 'Understanding political alignments in contemporary Britain: do localities matter?', *Political Geography Quarterly* 6: 53-76.

Sinclair, K. (1969) *A History of New Zealand*. Auckland: Penguin Books.

Taylor, P. J. and Gudgin, G. (1977) 'Antipodean demises of Labour', in R. J. Johnston (ed.) *People, Places and Votes: Essays on the Electoral Geography of Australia and New Zealand*. Armidale: University of New England.

—— and Johnston, R.J. (1979) *Geography of Elections*. Harmondsworth: Penguin.

Chapter seven

An ecological perspective on working-class political behaviour: neighbourhood and class formation in Sheffield

Munroe Eagles

Despite receiving a great deal of scholarly attention, a thorough understanding of the relationship between class and political behaviour in Britain remains elusive. One particularly challenging debate centres around what has come to be known as the 'ecological paradox' in class voting. The paradox arose when divergent conclusions regarding recent trends in class voting were reached using analyses based on public opinion survey data and others based on aggregate census and electoral data. Specifically, survey researchers have continually turned up evidence of a decline of class voting (Franklin, 1985, among others) while analysts using aggregate data at the constituency level have persistently documented either continuity or even a strengthening of the class alignment (Miller, 1978; Eagles and Erfle, 1988). To account for these discrepant results, William Miller (discoverer of this ecological paradox) argued that the nature of the class alignment is shifting such that it ' ... becomes less and less about "people like me", and more and more about "people around here"' (1978: 283).

Central to Miller's argument is his contention that class sentiments and politics are being influenced more by aspects of life in residential neighbourhoods than by an individual's position in the labour market or in a functionally defined class structure. This has proved to be a controversial position among students of British political behaviour. (One study, for example, concluded that Miller's interpretation was 'not only misleading but entirely false'; see Kelley and McAllister, 1985: 582–3.) Others, however, have been more enthusiastic (e.g. Scarborough, 1987: 237–40).

This chapter contends that the debates which Miller's provocative analysis touched off have been inconclusive largely as a consequence of limitations associated with available data (see also Books and Prysby, 1988: 224). Data from residents of three working class neighbourhoods in Sheffield are analysed to extend the empirical assessment (i.e. to provide something like a 'plausibility probe' in Eckstein's (1975: 108–13) terminology) of the ecological perspective. The first sections of the chapter outline the ecological perspective, describe the data base for this analysis as well as the extent of working class formation among the respondents. The bulk of the analysis, however, is directed toward three recurring controversies in research from the ecological perspective. Explored first is a disagreement among scholars concerning whether sentiments of neighbourhood attachment help or hinder the processes of working class formation. Second, two alternative explanatory mechanisms thought to be responsible for the production of local effects on the behaviour of individuals are tested against the data. Finally, data bearing on the motivational assumptions responsible for conformity to group norms, hypothesized as underlying both interpretations of contextual influence processes, are presented.

Clear-cut answers to such complex questions are not forthcoming from the data analysis. In particular, a full and more rigorous test of the ecological perspective will assess its explanatory power in comparison with that of plausible rival perspectives. Given the absence of basic information concerning fundamentally important issues confronting the ecological perspective, however, it is worth while to deal first with the evidence pertaining to this perspective alone. The patterns identified in this analysis serve to raise questions concerning the plausibility of the most popular conceptualization of the ecological perspective (which depicts it as operating through social or psychological processes based on the local neighbourhood).

The ecological perspective on class formation

The politicization of class has long been taken to be the *sine qua non* of modernity. Despite a strong research tradition that viewed the links between class position, class consciousness, and behaviour as relatively unproblematic and straightforward, there is a growing awareness that the relationships among these are far from determined. The study of class formation attempts to understand the process whereby individuals sharing some economic attribute or relationship become aware of their common situation and interests (collectively referred to here as class sentiments) and develop a capacity to act collectively in support of these interests (class-based political behaviour). Class formation, to borrow Marx's oft-quoted terminology, refers to the process whereby a 'class in itself' is transformed into a 'class for itself' (Marx, 1847: 160; see Przeworski, 1977).

The defining characteristic of the ecological perspective on working class formation is its contention that class members are influenced in their awareness, identification, and consciousness of this class position, as well as in their propensity to act politically on the basis of their class position, by some feature of their residential social context. In his analysis of class voting in Middletown (USA), Thomas Guterbock articulated this perspective: 'If we mean by social class a person's overall station in life and a concomitant outlook, then we probably get a truer notion of a Middletowner's class position from a street address than we do from a job title' (1980: 1053–4; see also Segal and Meyer, 1969).

Residential context effects can be identified with respect to a wide range of dependent behavioural and attitudinal variables (race relations, public opinion on a variety of issues, political efficacy, participation rates, etc.) but there is especially good reason to expect spatial factors to influence the politicization of class. Class members are not distributed randomly across space. Rather, one of the most commonly noted processes of modern societies is the tendency of social classes to become segregated, resulting in patterns of residential differentiation on class lines (see Timms, 1970; Harvey, 1985, especially pp. 109–24; and Harris, 1984). As a consequence of these general segregation processes, most individuals will dwell among others of a similar station in life. Accordingly, sentiments of neighbourhood attachment and localism, as well as patterns of local social interaction, are likely to manifest an inherent class bias and therefore have a significant impact on the development of class-related sentiments and behaviour.

Evaluating the ecological perspective: the research setting and methods

The present study was designed with the aim of collecting data suitable for assessing the ecological perspective on class formation (Eulau and Rothenberg, 1986: 152–3; Scarbrough, 1987: 237, are among those recently calling for such studies). Specifically, this involves obtaining information about individuals, their social environments, and their perceptions of, and social interactions within, these environments. Sheffield was selected as an urban setting for the study for essentially two reasons. The primary theoretical interest of the study was in the process of working class formation and since Sheffield has been referred to as 'the most working class city in Western Europe' (Young, 1983: 85) it forms a near perfect setting for the study of working class politics at an advanced stage.

Furthermore, Sheffield's uneven topography has combined with class segregation processes to produce a diverse array of socially homogeneous residential communities (see Johnston and Rossiter, 1980: 848–9). Sheffield's working- class neighbourhoods, therefore, are pure and thereby constitute settings generally conducive to the kinds of processes thought to underlie the production of contextual effects on class voting. It is worth underlining, then, that the argument of this chapter is not that Sheffield is in any sense typical of a British norm; rather, it provides a context within which the political characteristics of interest are unusually visible, and within which the processes of theoretical concern (associated with class-homogeneous neighbour-hoods) ought to be particularly accessible.

Within Sheffield, three small areas were selected to represent distinctive neighbourhoods. Previous research suggested that the relevant spatial unit for the operation of contextual influence processes (often referred to in British literature as 'neighbourhood effects'; see Busteed, 1975: 46; Johnston and Taylor, 1979, especially chapter five, pp. 221–69) is the local neighbourhood. In turn, research on neighbourhood perception in England, particularly in large cities such as Sheffield and among working-class individuals, has emphasized the extremely small scale of these units. William Hampton's definitive survey-based study of neighbourhood sentiments in Sheffield found that nearly three-quarters of the residents sampled described their home area as the few streets surrounding their address, and 7 per cent claimed to be at home only in their street (Hampton, 1970: 106).

Accordingly, the areas chosen to represent neighbourhood units for this research were comprised of several streets, each containing roughly 250 to 400 homes. In an effort to incorporate something of the diversity of working-class neighbourhoods, the areas selected were differentiated primarily with respect to the predominant form of housing tenure.[1] Of the 877 questionnaires hand- delivered to each household during April and May 1986, completed questionnaires were received from 389 (an overall response rate of 44.4 per cent, drawn relatively evenly from the three neighbourhoods). Considering the small size and homogeneity of these areas, the resulting data set provides a relatively solid foundation for the analysis of the politics of all residents.

Class position and class formation: general patterns

Social class, it is generally acknowledged, is a complex multi-dimensional concept. The preferred operationalization of class has been in terms of an individual's occupation (Reid, 1977: 12–22; Johnston, 1985b: 29–30). This study adopts the class coding practices developed by Heath *et al.* (1985: 13–16) with one major qualification: whereas it is customary in survey research to assign 'class by association' to individuals out of employment at the time of interviewing (either by reference to the occupation of the head of their household, or by reference to their previous occupation), this study attempts to avoid potential biases resulting from this procedure (see Peake, 1984; Dunlevy and Husbands, 1985: 124–9) by differentiating further between those employed and not employed outside the home. Nationally, non-employed individuals (i.e. the retired, students, housewives or unemployed) may comprise over a third of survey respondents (e.g. Miller, 1978: 258). Among some sub-groups of society (notably women, or those living in economically depressed areas, for example) the figures are likely much higher. In Sheffield, for example, by 1985 people with jobs were outnumbered by those without employment (Sheffield CPU, 1986: 31). These individuals, who occupy a dubious position in the class structure, may develop interests distinct from those in employment (and which may not be well represented by traditional coding practices).

Table 7.1 provides several pieces of information on the class composition and levels of working-class formation found in the study areas. Occupational class positions could be assigned to 57.3 per cent of the respondents, while a further 38.1 per cent were non-employed. With over half of all those in employment at the time of the study working in rank and file manual (i.e. proletarian) jobs, the heavily working class character of the Sheffield neighbourhoods is readily apparent.

Table 7.1: General indicators of working class formation among residents of Sheffield neighbourhoods (%; N in brackets)

	Salariat	*Intermediate*	*Proletariat*	*Non-employed*
Proportion of	9.7	18.2	32.3	39.9
respondents	(36)	(67)	(120)	(148)
Proportion claiming				
a class	69.4	71.6	86.6	78.4
identification	(25)	(48)	(103)	(105)
Of these, proportion	52.2	80.4	95.0	88.1
working class[*]	(12)	(37)	(94)	(96)
Level of working class[+] consciousness				
Low	61.9	46.5	21.2	23.3 (74)
Medium	28.6	27.9	47.5	34.9 (95)
High	9.5	25.6	31.3	41.9 (80)
Percentage Voting	44.4	52.7	82.9	78.2
Labour, 1983	(12)	(29)	(87)	(93)
Proportion highly				
loyal[+] to	30.6	22.4	40.8	45.3
Labour Party	(11)	(15)	(49)	(67)

[*] Heath *et al.*'s 'working class' category, renamed here following Crewe (1986): 624.
[+] For details of index construction see Appendix 7.1

How closely are these class positions related to variations in class sentiments? A baseline indication of the nature of class sentiments generally, and a minimal condition of class formation, is the presence of a subjective class identification. As David Robertson has remarked on the basis of an analysis of spontaneous class identifiers (i.e. respondents who are immediately willing to provide a class identification when asked by an interviewer) in Britain during the 1964–79 period, ' ... merely giving oneself a class label, not more common than 50:50 on probability, is to mark oneself out as unusually aware of social class' (1984: 30). Questionnaire responses to a question of class identification are necessarily spontaneous since there is no opportunity to prompt reluctant respondents to identify with a class.

By this standard, the Sheffield respondents are considerably more class aware than is the British norm. Looking at the second line of Table 7.1, over three-quarters of all respondents (76.8 per cent) indicated that they thought of themselves as being a member of a social class. When subjective class identifiers were asked which class they identified with, 85.1 per cent claimed to be 'working-class'. Moreover, a majority in each class division, including those in salaried occupations, identified with this class. Although levels of working class identification are highest among proletarians (95 per cent of those who claimed a class identification), over two thirds (67.1 per cent) of *non*-proletarians in employment claimed also to be working class. In other words, the homogeneity in the class identifications of residents evident in Table 7.1 is strikingly consistent with expectations of neighbourhood homo-geneity based on the ecological perspective.

To assess further the level of working class formation among residents, responses to four questions were combined to form an additive index of 'working class consciousness' (See Appendix 7.1 for details of the construction of this and other indices.) Table 7.1 also presents the cross-classification of class position and the level of working-class consciousness. Once again the heavily working class character of the respondents is apparent. Even roughly one in ten respondents with salaried occupations, and a quarter of members of the intermediate classes, scored high on *working*-class consciousness. Proletarians were only slightly more likely to evince high levels of working-class consciousness than these other two classes, and were less likely to do so than the non-employed. On the measure of class consciousness also, there is evidence suggestive of a contextual effect such that residents of working-class neighbourhoods were prone to adopt the outlook of the locally dominant working class, often without regard to their personal class position.

The relative homogeneity of working class formation among residents extends to the realm of voting behaviour, and the Sheffield residents certainly bear out the city's reputation as a Labour stronghold. In all, 72 per cent of respondents reported voting Labour in 1983, a time when Labour's national support fell to a post-war low of 28 per cent.

While over eight in ten proletarians reported voting Labour (roughly the same proportion as the non-employed), even 44 per cent of salaried workers did so also. To provide a more refined indication of the intensity of Labour support among members of the different classes, an index of 'Labour loyalty' was computed (see Appendix 7.1). The final row of Table 7.1, showing only the proportions exhibiting high levels of Labour loyalty, provides even further confirmation of the solidly Labour character of these neighbourhoods. Salaried workers (those expected to be least loyal) were slightly more likely to evince high levels of Labour loyalty than were members of the intermediate classes, and most loyal of all were the non-employed, of whom 45.3 per cent were loyal Labour partisans. Among proletarians, by comparison, the proportion of highly loyal Labour supporters was 40.8 per cent.

These results dramatically confirm the overwhelmingly working class ethos of the three study areas. They also illustrate convincingly that there is more to working class formation than the mere holding of a typically working class occupation. Many respondents whose personal class position did not place them unambiguously in the rank and file of the working class (i.e. the proletariat) also manifested attributes characteristic of reasonably advanced levels of working-class formation. In sum, though objective class differences in the levels of working class formation exist, they appear muted. Instead, the relatively high incidence of typically working class orientations, contributing to high levels of local political homogeneity in the three study neighbourhoods, is evident. To what degree can this be attributed to social and/or psychological processes identified by proponents of the ecological perspective? The following section explores the three issues facing the ecological perspective mentioned in the introduction.

Ecological influences on class formation: causes and consequences

First, is there any evidence to the effect that respondents for whom the residential neighbourhood constituted a particularly meaningful or positive environment were distinctive in terms of their class sentiments or political behaviour? In other words, do social solidarities based on the residential neighbourhood help or impede the process of working class formation? Social historians and political analysts are divided on this issue (see Cannadine, 1982; Lloyd n.d.: 63–76; Harris, 1984), and yet surely no more fundamental question confronts the ecological perspective on working class formation.

Neighbourhood orientations and class formation

To answer this question, four indicators of the respondents' relationship to their residential environment were constructed. First, respondents were distinguished according to whether they felt that 'there was an area around their home which they belonged to or where they felt at home (i.e. a neighbourhood)'.[2] Responses to this question constitute an important baseline measure of an individual's feeling of attachment to his/her residential neighbourhood (Hampton, 1970; Janowitz and Kasarda, 1974).

Second, a multi-item measure of the salience of these neighbourhoods was constructed (see Appendix 7.1). Individuals who carry out a relatively large number of important life-functions within the confines of their neighbourhood will have more opportunities to interact with their neighbours than those residents whose life functions are performed in geographically more extensive areas. They may also experience, in part perhaps as a consequence of this, additional incentive to conform with the local ethos (Eulau and Seigal, 1980: 22). Third, a composite measure of 'residential satisfaction' was constructed in order to gauge the effective dimension of an individual's orientation to neighbourhood. Finally, an index of 'neighbourhood integration' was constructed to measure the degree to which individuals felt psychologically and emotionally embedded in their residential environment. In the cases of the latter three indices, high scores denote positive orientations to neighbourhood.

These four orientations to neighbourhood tap various aspects of the nature of residential solidarities that, according to the ecological perspective, ought to be of significance to the process of working class formation. As noted above, scholars differ as to whether these extreme orientations to neighbourhood should be associated with specifically high or low scores on the various indicators of working-class formation. However, since the findings of most aggregate and contextual analyses suggest that working-class formation is enhanced by class segregation (e.g. Miller, 1978; Stephens, 1981: 164 etc.), the specific hypothesis tested here is that positive orientations to neighbourhood ought to be associated with higher levels of working-class formation, regardless of personal class position.

Table 7.2 presents in summary form the results of a series of cross-classifications illustrating the impact on the class position/class formation relationship of extreme (positive or negative) orientations to neighbourhood. To simplify and condense the presentation of this test, the proportion of respondents in the low category of the orientation to neighbourhood variable were subtracted from those in the high, and these differences alone appear in the table. Positive differences confirm that residential solidarities and working-class formation are complementary phenomena, and vice versa. An even split (or generally small differences) suggest that residential solidarities are of little consequence for the processes of working class formation.

Table 7.2: Impact of neighbourhood orientations on measures of class formation (percentage differences on dependent variable in the High category of neighbourhood orientation minus percentage in Low)

	Class position		
	Non-proletarians	*Proletarians*	*Non-employed*
Working-class identification			
Identification (Y–N)	20.1	–0.7	–19.0
Salience (H–L)	25.4	5.2	–2.1
Satisfaction (H–L)	19.0	–6.2	12.5
Integration (H–L)	29.2	6.5	10.0
High level of working-class consciousness			
Identification (Y–N)	12.7	–3.0	–17.8
Salience (H–L)	18.5	19.7	6.6
Satisfaction (H–L)	11.9	–11.6	30.2
Integration (H–L)	15.1	4.2	39.0
Labour voting, 1983			
Identification (Y–N)	19.8	–1.7	2.5
Salience (H–L)	8.4	–1.6	–5.9
Satisfaction (H–L)	5.9	–3.1	–6.8
Integration (H–L)	6.5	9.7	36.2
High levels of Labour loyalty			
Identification (Y–N)	23.6	36.5	5.2
Salience (H–L)	12.1	–17.7	–3.8
Satisfaction (H–L)	–9.0	12.2	–15.5
Integration (H–L	21.5	–9.3	1.0

* Various orientations to neighbourhood outlined in text and Appendix 7.1

The results suggest that a simple characterization of the relationship between orientations to neighbourhood and class formation in working class neighbourhoods would be misleading. Over a third of the forty-eight relationships reported in Table 7.2 are negative (meaning that those with the most positive orientations to neighbourhood were less likely than those with the least positive orientations to possess the characteristics of working class formation) and hence run counter to the hypothesis being tested. The majority (roughly two-thirds of the relationships reported) of positive differences, however, suggest that in most cases the two sources of collective sentiment (class and residential community) most often reinforce one another. Clearly, however, the contribution of neighbourhood orientations and solidarities to class formation is neither uniform nor (gauging from the magnitude of differences) dramatic.

By looking at the patterns within different class groupings this judgement can be refined. Of the sixteen relationships for non-proletarians, fifteen (94 per cent) are positive and hence in line with the complementarity hypothesis being tested. The proportion of positive differences falls to 62.5 per cent for the non-employed and falls further to 43.8 per cent for the proletarian respondents. Apparently, for the latter groups, residential solidarities do not correspond with those of class. For those whose personal class position is least congruent with that of the local neighbourhood, positive orientations toward that neighbourhood are associated with generally higher levels of working class formation.

Contextual influence processes and class formation

The second broad question to be addressed concerns uncertainty over the mechanisms responsible for the production of a homogeneous local socio-political ethos. Do residents modify their attitudes and behaviour as a result of impressions they form of their residential environment (perception-mediated processes), or do they arrive at their attitudes and behaviour intersubjectively, as a result of social interaction with their neighbours (interaction- mediated processes)? To assess the plausibility first of perception-mediated influence processes, Table 7.3 presents the proportions of respondents possessing the various indicators of working class formation in each class according to whether or not they perceived either a locally dominant class or partisanship. If perception-mediated contextual influence processes are responsible for the political homogeneity of these three neighbourhoods, those reporting such perceptions ought to be more likely to possess these indicators of working class formation than others.[3]

The data reveal some supportive evidence for the perception-mediated interpretations of contextual influence. Those perceiving a locally dominant class (with one minor exception in the case of the proportion of highly loyal Labour partisans among proletarians) were more likely to exhibit characteristics of both stages of working class formation. Similarly, perceiving a locally dominant party also was associated both with heavier Labour support in 1983, and with high levels of Labour loyalty, suggesting that perhaps impressions of *both* the social *and* the political environment are of behavioural relevance. Unlike the analysis of orientations to neighbourhood, where class differences emerged, the patterns of perceptual impact on class formation did not vary markedly across the class categories.

Table 7.3: Class position and perception-mediated influences on class formation (%; N in brackets)

	Local class			Local party		
	Yes	No	Difference (Y–N)	Yes	No	Difference (Y–N)
Non-proletarians (employed)						
Working-class identity	82.9 (34)	53.6 (15)	29.3			
High working-class consciousness	26.3 (10)	11.5 (4)	14.8			
Labour 1983	44.1 (26)	34.1 (15)	10.0	48.5 (16)	38.5 (25)	10.0
High Labour loyalty	33.3 (16)	27.3 (9)	6.0	48.0 (12)	25.5 (13)	22.5
Proletarians						
Working-class identity	100.0 (70)	77.4 (24)	22.6			
High working-class consciousness	32.9 (23)	27.6 (9)	5.3			
Labour 1983	73.8 (59)	70.0 (28)	3.8	86.1 (37)	64.3 (45)	21.8
High Labour loyalty	42.3 (30)	50.0 (19)	-7.7	46.3 (19)	42.9 (27)	3.4
Non-employed						
Working-class identity	91.5 (54)	86.4 (38)	5.1			
Working-class consciousness	43.4 (23)	39.4 (13)	4.0			
Labour 1983	66.2 (51)	59.7 (40)	6.5	73.9 (34)	57.8 (48)	16.1
High Labour loyalty	54.0 (34)	52.9 (27)	1.1	63.2 (24)	50.0 (32)	12.2

To this point the results are supportive of the perception-mediated hypotheses drawn from the ecological perspective. A closer look at the magnitude of the differences associated with the perception/non-perception of local characteristics quickly dampens any enthusiasm that may have been building. In general, those forming such impressions of the social composition of their residential environment are not favoured with dramatically higher levels of class formation than others without such perceptions. Moreover, many respondents did not form such environmental perceptions in any case (34.7 per cent did not recognize a locally dominant class; 66.3 per cent did not recognize a local partisanship), and hence were not liable to have been influenced by these perceptual influence processes.

Can differences in local social interaction provide a better account of the variations in class formation among residents? To answer this question it is useful to distinguish casual social interaction among co-residents from interactions with more explicit political content. Casual interaction in the neighbourhood setting may have political impact, according to Robert Huckfeldt, ' ... not because it is intensive or intimate, but rather because it is frequent and recurring: *the most important social learnings are those that are reinforced continually*' (1986: 130, emphasis in original; see also Weatherford, 1982: 130). Chance conversations, whether occurring, for example, while queuing at the shop or for a bus, or while relaxing in the local pub, may not have explicitly political content. Yet it is likely that such exchanges will reflect and embody the common concerns arising from the similar circumstances of residence in the same neighbourhood. Given the widespread class segregation characteristic of modern urban society even such casual social interaction may reinforce class- related sentiments and behaviour.

Measuring such casual interaction directly is difficult. One plausible surrogate has, however, already been introduced. Scores on the orientation of neighbourhood salience represent roughly the frequency of a respondent's opportunities for such casual interaction and are in part a consequence of the amount of time and number of activities an individual carries out in the neighbourhood. We have already seen that this factor (along with the other orientations to neighbourhood) appears to influence the level of class formation primarily among employed *non*-proletarians. Measures of salience may represent the opportunity for casual interaction, but they do not reveal how many opportunities will be taken advantage of by residents, nor do they give any indication of the general tenor of those interactions. A better measure of these latter qualities is the respondent's impression of the friendliness of other residents. Those finding other residents generally cordial will be more likely to take advantage of such casual conversational opportunities as come their way than others, and perhaps more prone to be influenced by them.

Looking for evidence of this among the Sheffield respondents, the data presented in the left-hand column of Table 7.4 suggest that there is some evidence for the strengthening of patterns of class formation among those finding other residents 'very friendly'. Once again, however, the differences between these respondents and others are neither consistent (a third run counter to expectations) nor particularly large. Taken together with the evidence on the impact of neighbourhood salience from Table 7.2, casual social interaction does not appear to have a particularly strong relationship with the indices of class formation.

Table 7.4: Class position, local social interaction, and class formation (%; N in brackets)

	Residents very friendly			Discuss politics with residents		
			Difference			Difference
	Yes	No	(Y–N)	Yes	No	(Y–N)
	Non-proletarians (employed)					
Working-class	80.0	65.9	14.1	100.0	65.2	34.5
identity	(20)	(29)		(11)	(20)	
High working-class	34.8	12.2	22.6	30.0	18.5	11.5
consciousness	(8)	(5)		(3)	(10)	
Labour 1983	46.9	52.0	−5.1	69.2	46.4	22.8
	(15)	(26)		(9)	(32)	
High Labour loyalty	27.5	24.2	3.3	50.0	21.4	28.6
	(11)	(15)		(7)	(19)	
	Proletarians					
Working-class	97.4	93.2	4.2	96.7	94.7	2.0
identity	(37)	(55)		(22)	(71)	
High working-class	32.5	31.6	0.9	37.5	29.7	7.8
consciousness	(13)	(18)		(9)	(22)	
Labour 1983	86.1	79.7	6.4	89.3	80.3	9.0
	(37)	(47)		(25)	(65)	
High Labour loyalty	37.5	44.9	−7.4	48.3	37.8	10.5
	Non-employed					
Working-class	91.9	89.7	1.4	95.7	89.0	6.7
identity	(41)	(52)		(22)	(73)	
High working-class	54.1	32.0	22.1	57.1	37.3	19.8
consciousness	(20)	(16)		(12)	(25)	
Labour 1983	76.9	78.1	−1.2	82.1	76.1	6.0
	(40)	(50)		(23)	(67)	
High Labour loyalty	40.6	48.2	−7.6	51.6	43.9	7.7
	(26)	(39)		(16)	(50)	

Table 7.4 also presents (in the right-hand columns) evidence concerning the impact on class formation of neighbourhood social interaction that has more explicitly political content. Specifically, it compares the incidence of various measures of class formation between those who reported discussing politics with their neighbours and those who did not. Here the results provide slightly stronger support for the existence of influence-mediated contextual influence processes. In all

three class categories on each indicator of class formation, those who reported political discussions with neighbours were more likely than others who did not to possess characteristics of working-class formation. In the case of non-proletarians in particular, the differences between political discussants and non- discussants was considerable.

Evidence from the Sheffield respondents suggests, therefore, that perception- and interaction-mediated contextual influence processes are likely to be complementary rather than mutually exclusive influences on the class formation process. However, neither version appears from the data analysis to offer an especially compelling account of variations in the measures of class formation. In other words, differences on measures of class formation of those who were implicated in these contextual influence processes and of others who were not, were generally smaller than would be anticipated if these mechanisms were largely responsible for the creation of local socio-political homogeneity. Apparently there is more to the homogeneity of these areas than can be accounted for by these representations of contextual influence processes.

The relative salience of neighbours

Underlying both the perception- and interaction-mediated explanations of contextual influence processes is the assumption that individuals will want to bring their own attitudes and behaviour in line with local sentiment. This assumption rests on the implicit judgement that of all potential social groups to which the individual might belong and from which he/she might receive social or political influence, neighbours will be the most salient (or at least, not be less salient). Otherwise, individuals might base their socio-political orientations either on their perceptions of other (non-local) social groups, or on the basis of interactions with members of these other groups. In other words, implicit in contextual explanation is an assumed hierarchy of group influences in which neighbours are at or near the top. Testing this assumption raises complex and thorny empirical problems. However, to complete the plausibility probe, some attempt to explore for evidence of the relative salience of neighbours should be made.

To accomplish this, responses to a battery of questions concerning the amount that respondents' felt they held in common with five different potential social groups were analysed. In addition to neighbours, respondents were asked to specify how much ('a lot', 'some things', 'little', or 'nothing') they felt they had in common with their workmates, the working class, the middle class, and others in their tenure group (i.e. other renters or owner-occupiers). The latter was included in order to test for evidence of 'housing classes' or cleavages

113

based on the consumption of housing. Comparing responses to these identical questions (arranged sequentially to form a grid on the questionnaire in order to encourage respondents to make their judgements of each group in a comparative context) will demonstrate whether the assumed hierarchy of group influences is actually in existence among the respondents. It is important to emphasize that comparisons made between groups were *not* supplied by the respondent, but rather were imposed analytically by comparing answers to identical but distinct questions regarding each group. The assumption in this test is that the amount of common interest felt with different groups will reflect the closeness of a respondent's sense of solidarity or belongingness in the group. If the calculation of common interests underlies the motivation for conformity with co-residents, then neighbours ought to be equally, if not more, salient than other potentially influential social groups.

Table 7.5: The relative salience of neighbours as against other social groups, by class

	More in common	Equal amount	Less in Common	Indifference Ratio[*]	Ratio[+] More/Less	(N)
Non-proletarians (unemployed)						
Middle class	10.5	50.5	38.9	.98	.27	(95)
Tenure group	13.8	61.7	24.5	.62	.57	(94)
Workmates	9.4	38.5	52.1	1.60	.18	(96)
Working-class	3.2	42.6	54.3	1.35	.06	(94)
Proletarians						
Middle class	43.5	41.7	14.8	1.40	2.94	(108)
Tenure group	30.7	63.4	5.9	.58	5.17	(101)
Workmates	4.6	66.4	29.1	.51	.16	(110)
Working class	6.4	43.6	50.0	1.29	.13	(110)
Non-employed						
Middle class	48.1	40.3	11.7	1.48	4.10	(77)
Tenure group	23.4	63.6	5.6	.45	4.50	(71)
Working class	6.0	63.9	30.1	.57	.20	(83)

[*] Indifference ratio = Number differentiating between neighbours and group / Number not differentiating between neighbours and group

[+] More/less ratio = Number having more in common with neighbours than group / Number having less in common with neighbours than group

Table 7.5 presents several pieces of information exploring the relative salience of neighbours among respondents in the different class groups. The three left-hand columns of the sub-tables for each class present the row percentages comparing responses to the amount felt to be had in common with neighbours as opposed to each of the other groups (more, equal amounts, or less). What is perhaps most striking in these columns is the high proportion of respondents in each class who decline to distinguish between the amounts they felt in common with neighbours and the other groups. This proportion is always greater than a third of all respondents, and in a number of cases it approaches two-thirds.

To provide a summary indicator of the propensity to differentiate between each group and neighbours, an indifference ratio was calculated (by dividing the number of individuals rating the amount felt in common with neighbours and the group in question differently by the number rating the two groups equally) from these figures and is presented as a fourth column. Ratios above 1 indicate that a higher proportion differentiated between the amounts held in common with neighbours and the group in question, while ratios between 0 and 1 denote a general disinclination to view common interests with both groups differently (and a ratio of 1, of course, indicates an even split on this). The fact that six of the eleven indifference ratios are less than 1 suggests that for most respondents neighbours are about as salient as other social groups.

Though the lack of strong polarization in responses to these questions is perhaps the dominant impression created by the data in Table 7.5, some clear and noteworthy class differences in the relative salience of neighbours as against particular groups emerge. Non-proletarians, for example, are far more likely to rate neighbours as lower (in terms of common interest) than the middle class or their tenure group than either proletarians or the non-employed. To help summarize the predominant direction of differentiation among those rating neighbours and other groups differently (those rating both equally are ignored in this aspect of the analysis), a second ratio (presented in column 5) was calculated in which the number rating neighbours higher than the group is divided by those rating them lower. Ratios between 0 and 1 indicate that a majority of those differentiating did so in favour of the other group; while ratios greater than 1 indicate the opposite (and 1 denoting equiprobability for either rating).

Looking at these ratios, non-proletarians tended to regard the amount of common interest with neighbours as less than that with all other social groups more frequently than not. Perhaps surprisingly, this was particularly the case when the other group was 'the working class' (ratio: 0.06) and 'workmates' (ratio: 0.18). Proletarians who differentiated tended to follow the same pattern when the comparison group was their workmates or the working class (ratios of 0.13 and 0.16 respectively), but they were far less likely than non-proletarians to rate the middle class or others in their tenure group higher than neighbours (ratios of 2.94 and 5.17 respectively). Responses of the non- employed were broadly comparable to the proletarians in the sample.

Summarizing the implications of this analysis for the ecological perspective, then, the general disinclination to differentiate among the amounts of common interest held with neighbours and other groups suggests that although there is little evidence to refute the self-interested motivational assumptions underlying the hypothesized desire for conformity (i.e. neighbours were for the most part not *less* salient than other groups), neither is there much positive support for it (i.e. neighbours were rarely considered to be more salient than other groups). For many, it seems that neighbours were strategically close to all other social groups mentioned in terms of the amounts of common interest. However, among those who did make distinctions between the amounts held in common with neighbours and other groups, when the comparisons involve the working class or workmates (for those employed) they were almost always unfavourable for neighbours. This suggests, therefore, that motivated conformity to group norms, to the extent that these are based on a calculation of the amount of common interest and to the extent to which they are important influences on class formation, are likely to involve groups with a more explicit working class composition than neighbours.

Conclusions

In terms of the plausibility of the ecological perspective, at least in its most popular conceptualization as a consequence of neighbourhood-level processes, the analysis can perhaps best be characterized as 'damning with faint praise'. Neighbourhood influences are detectable, but in themselves they seem inadequate as explanations for variations in the indicators of class formation manifested among the respondents. This is not to suggest that context is unimportant as an explanation of the political behaviour of these respondents, but rather that those 'influences' operating at the neighbourhood level, despite the literature on neighbourhood effects which suggests otherwise, are of limited significance. Contextual influences broader than the neighbourhood cannot be ruled out. Political homogeneity may spring from a political environment broader than the residential neighbourhood. After all, respondents in this study were residents of Sheffield, a city with a proud and strong Labour tradition (and itself part of the 'Socialist Republic of South Yorkshire').

In this respect, failing to account for the relatively high levels of working class formation using neighbourhood factors adds credence to some of the recent arguments of R. J. Johnston, an early proponent of the study of neighbourhood effects. Johnston contends (and this research supports his contention) that a neighbourhood effects explanation of local political homogeneity presents a false impression of the factors creating local political cultures ' ... because its ahistorical representation of social processes ignores the local institutional and cultural milieux within which people are socialized' (1985a: 252–3; 1986: 113–15; see also Agnew, 1987). In advocating an interpretation of local homogeneity that is set 'more firmly into the context of place-based socialization' (Johnston, 1985a: 253), a major step towards narrowing the gap separating critics (who also emphasize the importance of socialization variables; see Kelley and McAllister, 1985; McAllister, 1987) and proponents of contextual analyses and the ecological perspective may have been taken.

Appendix 7.1 Scoring and index construction

Following are details on the scoring and construction of the multi-item indices employed in the chapter, presented in the order in which they appear in the text.

Working class consciousness. Respondents were given a score of 1 for the purpose of constructing this index if they: identified with the working class; felt closer to members of their own class than people of other classes; felt they had 'a lot' in common with other members of the working class; and felt that there is bound to be some conflict between social classes. Otherwise, respondents were given zeroes on these index components. The resulting five-point index (possible scores ranged from 0 to 4) was collapsed into three categories representing high, medium, and low levels of working class consciousness (by grouping scores of 0 and 1 for the low and 3 and 4 for the high categories, and letting scores of 2 represent a medium category).

Labour loyalty. Respondents were assigned a score of one for each of the following; if they expressed a Labour Party identification; indicated that they felt this identification 'fairly' or 'very' strongly; would vote Labour if an election were called then; and if they claimed always to have voted for Labour. Otherwise, respondents were given zeroes on these items. The five point index (0–4) formed by adding these responses together was collapsed into three categories (0, 1 = low; 2, 3 = medium, 4 = high levels of Labour loyalty).

Neighbourhood salience. Individuals who worked within a mile of their home, spent their leisure time in the neighbourhood, claimed to have a local pub, and shopped for day-to-day items in the neighbourhood were each given a score of 1 on this measure (otherwise they were assigned zero). Combining these formed a five-point index of salience (0–4), which was collapsed into three categories by combining the two extreme categories (0, 1 and 3, 4).

Residential satisfaction. Respondents received a score of 3 each time they responded with the most positive option (i.e. who chose 'very safe', 'very happy', or 'very sorry'), 2 when they selected the middle ('somewhat') category, and 1 when they indicated the least positive response ('not very'). Combining these scores produced a seven-point index of residential satisfaction for each individual, which was then collapsed into three categories (low = scores of 1-3; medium = scores of 4, 5; high = scores of 6, 7).

Neighbourhood integration. Component questions tapped the respondents' level of interest in the affairs of the district, their sense of how much they had in common with other residents, and whether or not they regarded the neighbourhood as their permanent home. Scores on the first two ranged from 3 (highest) to 1 (lowest), while the latter was scored 3 if yes, 2 if uncertain, and 1 if not.

Notes

1. The assistance of Professor William Hampton, University of Sheffield, in identifying suitable neighbourhoods is gratefully acknowledged. One area was located on a massive council estate built between the two world wars in the north of the city, a second was a tract of owner-occupied semi-detached homes in the eastern section of the city, and the third was composed entirely of turn-of-the-century terraced houses located in the city's southern sector. Roughly two-thirds of these small dwellings were being purchased by their occupants, with the rest being privately rented.

2. Those who responded negatively to this question were asked to answer other questions referring to their home area ' ... as if they referred to a few streets surrounding your home'. Otherwise, respondents were left completely free to define the size and boundaries of their home area. For a useful discussion of this measure, see Hampton (1970): 99-100.

3. A subsequent probe revealed that, of those perceiving a locally dominant class, 93 per cent indicated that this was 'the working class' while of those noting a locally dominant party, 78 per cent responded that it was Labour. Since individuals exhibiting high levels of class consciousness and politicization may also be more sensitive to the class character of their environment, the issue of causality is problematic here. Failing to find a particularly strong relationship here between class sentiments and neighbourhood perceptions suggests that the hypothesized relationship is at least plausible.

References

Agnew, J. A., (1987) *Place and Politics: the Geographical Mediation of State and Society*, London: George Allen & Unwin.

Books, J. and Prysby, C. (1988) 'Studying contextual effects on political behavior: a research inventory and agenda', *American Politics Quarterly* 16, 2: 211–38.

Busteed, M. A. (1975) *Geography and Voting Behaviour*, London: Oxford University Press.

Cannadine, D. (1982) 'Residential differentiation in nineteenth-century towns: from shapes on the ground to shapes in society', in J. H. Johnson and Colin G. Pooley (eds.) *The Structure of Nineteenth Century Cities*, London: Croom Helm, 235–51.

Crewe, I. (1986) 'On the death and resurrection of class voting: some comments on "How Britain votes"' *Political Studies* XXXIV: 620–38.

Dunleavy, P. and Husbands, C. T. (1985) *British Democracy at the Crossroads: Voting and Party Competition in the 1980s*, London: Allen and Unwin.

Eagles, M., and Erfle, S. (1988) 'Community cohesion and working class politics: workplace-residence separation and labour support, 1966–1983', *Political Geography Quarterly*, 7: 229–50.

Eckstein, H. (1975) 'Case study and theory in political science', in F. Greenstein and N. Polsby, eds., *Handbook of Political Science, VI: Strategies of Inquiry*, Reading, Mass.: Addison-Wesley, 79-137.

Eulau, H. and Rothenberg, L. (1986) 'Life space and social networks as political contexts', *Political Behavior* 8, 2: 130–57.

—— and Seigel, J. (1980) 'A post-facto experiment in contextual analysis: Of day- and night-dwellers', *Experimental Study of Politics* 7, 2: 1–26.

Franklin, M. N. (1985) *The Decline of Class Voting in Britain: Changes in the Basis of Electoral Choice, 1964-83*, Oxford: Clarendon Press.

Guterbock, T. (1980) 'Social class and voting choices in Middletown', *Social Forces* 58, 4: 1044-56.

Hampton, W. (1970) *Democracy and Community: a Study of Politics in Sheffield*, London: Oxford University Press.

Harris, R. (1984) 'Residential segregation and class formation in the capitalist city: a review and directions for research', *Progress in Human Geography* 8, 1: 26–49.

Harvey, D. (1985) *The Urbanization of Capital: Studies in the History and Theory of Capitalist Urbanization*, Baltimore, Md.: The Johns Hopkins University Press.

Heath, A. Jowell, R., and Curtice, J. K. (1985) *How Britain Votes*, Oxford: Pergamon Press.

Huckfeldt, R. H. (1986) *Politics in Context: Assimilation and Conflict in Urban Neighbourhoods*, New York: Agathon Press.

Janowitz, M., and Kasarda, J. (1974) 'The social construction of local communities', in T. Leggatt (ed.) *Sociological Theory and Survey Research: Institutional Change and Social Policy in Great Britain*, Beverley Hills, Cal.: Sage, 207–36.

Johnston, R. J. (1985a) 'Class and the geography of voting in England:

towards measurement and understanding', *Transactions, Institute of British Geographers* 10 2: 245–55.

—— (1985b) *The Geography of English Politics: The 1983 General Election*, London: Croom Helm.

—— (1986) 'Place and votes: the role of location in the creation of political attitudes', *Urban Geography* 7, 2: 103–17.

—— and Rossiter, D. J. (1980) 'Geography is the clue to election victories', *Geographical Magazine* 52: 848–9.

—— and Taylor, P. J. (1979) *The Geography of Elections*, London: Croom Helm.

Kelley, J. and McAllister, I (1985) 'Social context and electoral behavior in Britain', *American Journal of Political Science* 29, 3: 564–86.

Lloyd, J. (n.d.) 'Neighbourhood, community and class: the role of place in the formation of class consciousness', *Papers in Urban and Regional Studies* 3, Birmingham: Centre for Urban and Regional Studies, 63–76.

McAllister, I. (1987) 'Social context, turnout, and the vote: Australian and British comparisons', *Political Geography Quarterly* 6, 1: 17–30.

Marx, K. (1847) *The Poverty of Philosophy*, repr. 1975, Moscow: Progress.

Miller, W. L. (1978) 'Social class and party choice in England: a new analysis', *British Journal of Political Science* 3: 257–84.

Peake, L. J. (1984) 'How Sarlvik and Crewe fail to explain the Conservative victory of 1979 and electoral trends in the 1970s – review essay', *Political Geography Quarterly* 3, 2: 161–7.

Przeworski, A. (1977) 'Proletariat into class: the process of class formation from Karl Kautsky's "The Class Struggle" to recent controversies', Politics and Society 7, 4: 343–401. Reid, I. (1977) *Social Class Differences in Britain: a Sourcebook*, London: Open Books.

Robertson, D. (1984) *Class and the British Electorate*, Oxford: Blackwell.

Scarbrough, E. (1987) 'The British electorate twenty years on: electoral change and the election surveys', *British Journal of Political Science* 17, 2: 219–46.

Segal, D. R. and Marshall, W. M. (1969) 'The social context of political partisanship', in M. Dogan, and S. Rokkan, (eds.) *Quantitative Ecological Analysis in the Social Sciences*, Cambridge, Mass.: MIT Press.

Sheffield Central Policy Unit (1986) *Sheffield: Putting you in the Picture*, Sheffield City Council, Central Policy Unit.

Stephens, J. D. (1981) 'The changing Swedish electorate: class voting, contextual effects, and voter volatility', *Comparative Political Studies* 14, 2: 163–204.

Timms, D. W. G. (1971) *The Urban Mosaic: towards a Theory of Residential Differentiation*, Cambridge: Cambridge University Press.

Weatherford, M. S. (1982) 'Interpersonal networks and political behavior', *American Journal of Political Science* 26, 1: 117–43.

Young, A. (1983) *The Reselection of M.P.s*, London: Heinemann Educational Books, London.

Chapter eight

Lipset and Rokkan revisited: electoral cleavages, electoral geography, and electoral strategy in Great Britain

R. J. Johnston

The launch of the Social Democratic Party (SDP) in Great Britain in 1981 was heralded by the claim that it was going to 'break the mould' of British politics (Bradley, 1981) and create a new central consensus which would end the adversary politics of the Conservative and Labour parties (Finer, 1975) and restore stability and prosperity to the British economic and social scene. The 'success' of that venture is best illustrated by the two landslide Conservative victories in the general elections of 1983 and 1987 when the SDP, in Alliance with the Liberal Party, won approximately a quarter of the votes at each contest but less than 5 per cent of the seats in the House of Commons. After the second of these elections, an attempt to merge the SDP and the Liberals resulted in a fragmentation of both, which will undoubtedly be to the benefit of the Conservative and Labour Parties (as opinion polls clearly illustrated in 1988 and 1989).

This attempt to restructure British politics illustrates a major conclusion in Lipset and Rokkan's (1967) classic analysis of electoral behaviour in West European liberal democracies. As they put it:

> the party systems of the 1960s reflect, with few but significant exceptions, the cleavage structures of the 1920s ... the party alternatives, and in remarkably many cases the party organiz-ations, are older than the majorities of the national electorates. To most of the citizens of the West the currently active parties have been part of the political landscape since their childhood, or at least since they were first faced with the choice between alternative 'packages' on election day.
>
> (p. 50)

This organizational stability clearly implies a geographical stability, one illustrated for a number of countries (Johnston *et al.*, 1987). But that stability carries with it certain implications with regard to the operation of politics, implications which developments like the SDP in Great Britain are now challenging. We need to understand those challenges in order to appreciate contemporary electoral geographies, which is the purpose of the present chapter, with a general focus on cleavage change. A review of Lipset and Rokkan's model suggests that in certain situations the continuity that they identify is indeed likely to be long-lasting but that a change in circumstances can lead to cleavage failure. The situation in Great Britain in the 1980s is used to illustrate such failures.

The basic model: Lipset and Rokkan revisited

Lipset and Rokkan's (1967) essay was only marginally related to electoral geography, but its implications have been substantial and it has implicitly underpinned much recent work (as Taylor and Johnston, 1979, suggest). Their basic concern was understanding the party systems of western (basically European) liberal democracies, where political parties are major institutional channels for the structuring of information flows and conflicts: their existence 'has helped to stabilize the structure of a great number of nation-states' (p. 5). Following Parsons's (1959) general model of society, therefore, political parties are major sources of stability and order in inherently conflict-ridden societies.

But what sort of order; what sort of political parties? European liberal democracies vary substantially in their party systems. Lipset and Rokkan's analyses suggest four basic cleavages in such societies. The first two are the products of the *national revolution*, the process whereby the modern nation-state emerged. During this process, there were frequently conflicts between interest groups seeking to build a new, centralized state and other interest groups opposed either to the state itself or to its *raison d'être*. Hence you got:

> the conflict between the central nation-building culture and the increasing resistance of the ethnically, linguistically, or religiously distinct subject populations in the peripheries: the conflict between the centralizing, standardizing, and mobilizing Nation-State and the historically established corporate privileges of the Church.
>
> (p. 14).

From this came two cleavages: that between subject and dominant cultures (or centre v. periphery), and that between the church and state. Political parties may have been established to mobilize the enfranchised population on one side of either or both of these cleavages. If successful, their continued role in the socialization and mobilization of the electorate gave them an established position in the organization of politics within the country.

The second set of cleavages was related to the industrial revolution, the process whereby industrial capitalism was established as the dominant mode of production. Again, two separate cleavages were identified, reflecting

> the conflict between the landed interests and the rising class of industrial entrepreneurs: the conflict between owners and employers on the one side and tenants, labourers, and workers on the other.
>
> (p. 4).

The first resulted in a cleavage between the primary and the secondary economy (or town v. country); the second was between employers and employees (often termed the class cleavage.

Lipset and Rokkan were concerned with the following issues, among others.

1. *The pattern of cleavages in a country*. All cleavages, and the conflicts that they reflect, were parts of protest movements, usually against an elite, often a new one. But they varied quite substantially from country to country:

> Quite different types of protest alignments have tended to occur in fully mobilized national states. ... In one way or another they all express deeply felt convictions about the destiny and the mission of the nation, some quite inchoate, others highly systematized; and they all endeavor to develop networks of organizations to keep their supporters loyal to the cause.
>
> (p. 23).

The reason was largely differences in the timing and character of both revolutions (p. 34), and the particular contexts within which those revolutions occurred. The result is an eightfold categorization of 'alliance-opposition structures', the outcome of three dichotomous divisions. The first occurred after the Reformation, with either state control of a national church (as in England) or the state allied to the Roman Catholic church; the second occurred after the 'democratic revolution', when the first two were further subdivided according to the strength of the 'established' church; and the third came after the industrial revolution, with a division between commitment to landed interests and commitment to urban interests.

2. *The sequence of cleavage development*. The sequence of three stages outlined above generally occurred prior to a substantial widening of the franchise. As the franchise was extended, so cleavage structures changed, leading in some countries, as Great Britain, to the growth of parties with strong agrarian bases and in others to the growth of parties with clear territorial bases. Whatever the particular pattern, however, it

was substantially modified by what Lipset and Rokkan (1967) call the 'decisive thrust toward universal suffrage' (p. 47). Thus the subject v. dominant culture, church v. state, and town v. country cleavages were present first – in varying degrees of strength – and produced the geographical variety currently observed. Into the cleavage structure in each country was then placed the class conflict. Thus

> the interactions of the 'center-periphery', state-church, and land-industry cleavages tended to produce much more marked, and apparently much more stubborn, differences among the national party systems than any of the cleavages brought about through the rise of the working-class movements.
>
> (p. 46)

This does not imply that the class cleavage introduced a similar party structure, of similar relative strength, to each country, however. Lipset and Rokkan argue that their focus of explanation was not the emergence of a distinctive working-class movement at some stage (that was to be expected) but rather

> the strength and solidarity of any such movement, its capacity to mobilize the underprivileged classes for action and its ability to maintain unity in the face of the many forces making for division and fragmentation.
>
> (p. 46)

The result of these stages is a shift in the organization and relative strengths of political parties, away from the first three cleavages and towards the class cleavage. (Indeed, Lipset, 1970, presents it as such.) This suggests a geography of voting akin to the geography of uneven development that is characteristic of the industrial revolution and its aftermath. Comparing relatively backward areas to the more developed cores, therefore, Lipset and Rokkan (1967) argue that

> In the one case the decisive criterion of alignment is *commitment to the locality and its dominant culture*: you vote with your community and its leaders irrespective of your economic position. In the other the criterion is *commitment to a class and its collective interests*: you vote with others in the same position as yourself whatever their localities, and you are willing to do so even if this brings you into opposition with members of your community.
>
> (p. 13).

This is a crucial argument in much that has been done in the field of electoral geography in recent years. The implications are that analysts studying the geography of voting should find:

(a) At best a weak 'commitment to locality' component in 'developed countries' (as exemplified by Kirby and Taylor, 1976: Rose and Urwin, 1975, have presented cross-country comparisons on the extent of this regional factor, and McAllister and Rose, 1984, have argued its insignificance in Great Britain).

(b) A dominant class cleavage (perhaps accompanied by a cultural cleavage: Lipset, 1970, suggests that religion may form the basis for this), so that the geography of voting is readily predicted by the geography of class composition.

With regard to the latter, a great deal of work has emphasized the class cleavage. Much recent writing on voting in Great Britain implicitly accepts the above two conclusions, so much so that some of them hardly investigate spatial variations at all (see Johnston, 1986a, for a review). And yet, even Alford's (1963) classic showing Britain to have the deepest class cleavage of the countries he studied implied that occupational class alone was far from sufficient to account for the pattern of voting between parties (let alone its geography) and Lipset's (1984) discussion of 'Social conditions affecting left voting' suggested three factors that influence working-class voting for social change, which he linked to community size and plant size.

More detailed analyses of the geography of voting in Britain have shown that the strength of the class cleavage varies considerably from place to place (e.g. Johnston 1985a). The clearest finding is that both the Conservative Party and the Labour Party tend to get more votes than expected from the class composition of the electorate where their expectation is high, but less than expected where the expectation is low. Thus the country is spatially more polarized electorally than it is socially (Miller, 1977), and increasingly so (Miller, 1984; Johnston et al., 1988). The reasons for this are the subject of much debate, since the popular conception of a neighbourhood effect (introduced to geographers by Cox, 1969) is dubious with regard to the implied process of 'conversion by conversation' (Johnston, 1986b). Within the context of Lipset and Rokkan's model, however, these findings clearly suggest that people are more likely to vote with their class nationally if they live in a place where that class is politically dominant locally, and therefore influential in both the long-term processes of *political socialization* and the short-term activities of *political evaluation* (Johnston, 1987a).

3. *The stability of cleavages.* Lipset and Rokkan's (1967) conclusion that the 1960s' cleavages in West European liberal democracies were similar to those in place forty years earlier suggests not just stable political systems and stable electoral geographies (see Johnston, 1983, 1987b) but also a situation of stable governments. If voters are locked in to particular cleavages and party alignments, then the relative strength

of the parties will change little, with clear consequences for the allocation of elected power. Apart from anything else, this would seem to be in the interests of only some of the parties – those with an apparently permanent access to power – so that the others would seek to alter the cleavage structure (or new parties would seek to break the mould of the existing structure) in order to alter the access to power.

Political parties are presented by Lipset and Rokkan as necessary institutions for ordering societies and channelling conflicts. If, however, certain parties are bound to lose, because they represent the numerically inferior interest groups in a conflict, then over time their viability will be called into question, unless they can manipulate the cleavage structure so that they have a chance of victory. This manipulation would undoubtedly involve the geography of voting.

In Summary. Lipset and Rokkan's model raises a number of issues which suggest that it is not entirely satisfactory as it stands. In part this is because it is now twenty years old, and Harrop and Miller (1987) have suggested the addition of further stages to mark changes in the organization of capitalist societies and the political shifts that have followed. (Following the national revolution and the industrial revolution, according to their schema, came 'the growth of the states' – with which is associated the public sector/private sector political cleavage identified by Dunleavy, 1979 – and 'post-industrialism', which is linked to the growth of education, affluence, and post-material values that suggest political issues not firmly linked to the traditional party structures: Inglehart, 1977, 1981.) The issues that I want to address in the rest of this chapter are linked to the stability of governments and underlying cleavages, and hence to the stability of the electoral geography, in the context of recent changes.

Cleavages, governments and electoral geography

Most European countries have experienced several changes of government in recent decades, which to some extent counters the stability argument implicit in Lipset and Rokkan's model. It suggests that at least a small proportion of the electorate must alter their voting behaviour at least occasionally, in order for some parties to increase their share and others to decline. Such shifts in party preference may reflect shifts in the voters' own positions, as the social structure of their society changes – social mobility may move them from one side of a cleavage to another. Alternatively, the structure of society may alter because those newly entering the electorate differ in their political orientations from those leaving it. Undoubtedly both these shifts have occurred, but they would both imply a hardening of the party system in one direction and not a situation where party fortunes wax and wane.

For the latter to occur, there must be some volatility, some vote-switching between elections.

Many studies have shown the amount of inter-party movement in voter preferences. It is frequently argued, however, that volatility has increased in recent decades, as people become less attached to the traditional parties and are less inclined to vote habitually for the party they were socialized into supporting. This process of *dealignment* (see Dalton *et al.*, 1984; Crewe and Denver, 1985), is not in the parties' interests, however: they want a stable cleavage structure because of the guaranteed support that it brings, leaving them to compete for a small proportion of the electorate who are 'floating voters', and who are the key to electoral success. If everybody is a floating voter, then you get the rapidly-changing cleavage structures described as the politics of failure by Osei-Kwame and Taylor (1984; see also Taylor, 1986), which some commentators associate with the situation in the United States also (Harrop and Miller, 1987: 5, 32).

In understanding the interrelationships among parties, government, and electoral geography we need two classifications – of electoral systems and of parties. Each is presented as a dichotomy only, but the classes identified are ideal types which provide a simplified, but analytically useful, overview.

Electoral systems vary considerably in the ways in which they organize politics, even within the relatively restricted group of Western liberal democracies (what Harrop and Miller, 1987, call 'competitive electoral systems'). The basic division used here is between *single-member plurality systems* and *multi-member constituency systems*. The latter are usually associated with the achievement of (near) proportional representation, whereas the former are not: the former are usually associated with parliaments dominated by two parties, whereas the latter are believed to favour parliaments with many parties represented. Neither of those associations is a necessary consequence of the type of electoral system (as the essays in Bogdanor and Butler, 1983 show) but as statements of general tendencies they remain valid (as Rae's, 1971, analyses make clear).

Parties. Lipset and Rokkan's classification of cleavage systems identifies each political party with a particular section of the electorate – either a community (of any size) irrespective of its social structure or an interest group irrespective of its members' residential location. The party is assumed to promote their interests, but Lipset and Rokkan pay little attention to the political activity that is involved. The link between the party and its interest group is important to the development of cleavages, however, as exemplified by a simple classification into: *ideological parties*, those which are committed to a clear set of goals, and retain that commitment even if it is clear that they will obtain only minority support among the electorate; and *pragmatic parties*, which are much more flexible, and while allied with a certain set of interests and goals are prepared to modify their policies in order to win and retain power.

Combining these two classifications suggests several consequences. The first is that ideological parties are more likely to be successful over long periods in multi-member electoral systems, for two reasons. First, they are likely to win representation, provided their support is above a certain threshold, because of the PR element of those electoral systems. Secondly, they are possibly able to influence governments in such systems, because with many parties present and, in many cases, none with an absolute majority of the seats in parliament, either their presence in a coalition is needed or their support in key votes is required: thus they can win concessions from the pragmatic parties which seek to be as big as possible in order to dominate coalitions. (Size is not the only criterion of ability to dominate coalitions: see Johnston, 1982.)

The second consequence is that pragmatic parties are likely to dominate plurality electoral systems. This is a clear implication of the likelihood that such systems have two major political parties only, so that in seeking electoral victory (i.e. a majority of seats in the parliament) each has to try and win the support of a majority of the electorate. Downs' (1957) classic study provided the foundation for this conclusion, without reference to the division of the electorate into single- member territorial constituencies (see Johnston, 1979). He showed that the two parties would share a great deal of common ground in the modal section of the electorate – even if, as the proponents of the adversary politics model suggest (Finer, 1975), they retreated to left or right of that common ground once in office. The third consequence is the well known one that in plurality electoral systems, especially those with single-member constituencies, parties are advantaged if their supporters are spatially concentrated to some extent, and significantly disadvantaged if they are not. The extent of the advantage/disadvantage depends on the percentage of the vote that they win nationally (Gudgin and Taylor, 1979); small parties benefit from concentrated support, whereas large parties are disadvantaged by it.

Since most parties begin small, they are clearly advantaged then if they have strong support in certain communities. This support should not be transient, but should be deepened by a process of voter mobilization in which the party is not simply a contestant for votes at an occasional general election but is implicated, in a variety of ways and through a multiplicity of links, in many aspects of community life. In these ways, it creates a milieu into which new voters are socialized and provided with a continuing context for the interpretation of political cues, including the competing policy offerings of other parties. From that base, the party then needs to extend its influence into other communities, so as to build a substantial parliamentary representation. It needs a stable geography.

Geographical stability will always be under threat, however, as competing parties seek to 'colonize' areas – or at least to destroy the community solidarity so that their pragmatic offerings are not immediately rejected because they are not linked to the party that traditionally represents the area. This geographic competition may be no more than a slight tilting of the balance in an area first one way and then another, so that its underlying foundation remains solid. Occasionally, however, the geography may be substantially redrawn.

Lipset and Rokkan reapplied: Great Britain – a nation dividing?

A common simplistic view of British politics until the present decade sees them as dominated over 50 years by two parties – Conservative and Labour – whose support came from either side of a deep class cleavage. The detail was much more complex, despite Alford's (1963) portrayal of British voting behaviour as 'pure' class voting and Pulzer's (1967, p. 98) frequently-quoted claim that

> class is the basis of British party politics; all else is embellishment and detail

Butler and Stokes's (1969, 1974) detailed analyses proved differently, but in any case Pulzer (1967: 104) had recognized the impossibility of the deep cleavage

> Indeed 'pure' class voting, or anything approaching it, would put an end to the British party system.

For many decades, for example, the Conservative Party's major support came from the minority class, but it

> survived the democratization of the country by its ability to gain votes of the poor.

For much of the half-century between the decline of the Liberals in the 1920s and the onset of dealignment in the 1970s, the Conservative and Labour parties were very close on a range of issues, so that people could transfer their support from one to the other fairly readily, without a major switch in attitudes. The Labour Party had a clearer ideological base, expressed in its constitution, but for most of the period it was very much a pragmatic party, suffering swings against it (as in 1931) that were countered by major movements towards it (in 1945). Its core support was in the working class, particularly in the communities that it had mobilized in the decades around the turn of the century, where the major industrial trades unions provided a powerful ally and co-agent. But many working-class people traditionally voted Conservative – more so relatively than middle-class people who traditionally voted Labour – and only in the general elections of 1945, 1950 and 1951 did its share of the votes cast approach half.

In that period of two-party dominance, with pragmatic competition for votes (and realization by some Labour leaders that the ideological base was potentially damaging to Labour's quest for electoral power), the parties accepted a consensus position on many of the central elements of economic and social policy (including the welfare state). Hence, as Downs's (1957) model clearly shows, both could be elected by a population most of whom accepted that consensus too. It is widely argued that in such a context oppositions do not win elections, governments lose them. Governments with a good record tend to be re-elected – though often the quality of the record is a construction of party propaganda assisted by the media.

This concept of voting as a reward has been formalized by political scientists as 'retrospective voting' (Fiorina, 1981) and much has been written in recent years linking macro-economic indicators to voting intentions and voting behaviour. (For Britain, see, for example, Sanders *et al.*, 1987.) As the celebrated American political scientist V. O. Key (1966) put it, 'Voters may reject what they have known; or they may approve what they have known. They are not likely to be attracted in great numbers by promises of the unknown.' Thus it could be argued that the Conservative Party was rejected in 1964 because voters were increasingly dissatisfied with its performance in power, and saw the Labour Party as a credible alternative. Similarly, the Labour Party was rejected in 1979 on the basis of negative slogans such as 'Labour isn't working' and 'Winter of discontent' used by a revived Conservative party which was seen as a viable alternative government. In 1983 and 1987 similar slogans were used by Labour against the Tories, but while there is evidence of some dissatisfaction with the Conservative record and ideology (Curtice, 1988), neither the Labour leadership (especially in 1983) nor its policies were seen as sufficiently attractive by enough voters for the party to be returned to power: many people dissatisfied with the Tory performance voted Alliance, and thereby split the opposition.

In such a situation parties need those retrospective evaluations, otherwise governments would not be unseated. But they also need a core of supporters who are unlikely to desert them under normal circumstances. The latter are necessary to give the party security and continuity; the former are necessary to give it the needed majority if its core support is less than 50 per cent of the electorate. Thus parties need to mobilize and to cultivate, to develop a hard core of substantial support. Without that, it would be very difficult for them when in opposition to present themselves as a viable alternative to the party in power. In part they win that loyalty through mobilization around an ideology; in part they do it by rewarding their core of supporters when they are in power, by serving their interests and getting favourable, to some extent long-term, retrospective evaluations.

When they are in power, parties must deliver on their promises: they must advance the interests of those whom they want to keep as loyalists and also those floating voters who they do not want to return to an opposition party. For the former, the delivery of promises reinforces the long-term retrospective evaluations; for the latter, it is the short-term evaluations that are crucial, hoping that voters will reward the government for recent trends. Hence the 'electoral cycle' that many claim to identify, whereby a government spends to win votes in the period prior to an election, and introduces deflation soon after. Governments lose either when the pre-election boom fails to materialize, for reasons at least partly outside their control, or because of the triumph of long-term over short-term memories.

This situation dominated British politics in the 1950s and 1960s, but thereafter began to break up, as both the Conservative and the Labour Parties shifted from the central consensus under pressure from more 'extremist' internal elements. In Labour there has always been a conflict between the ideological, pro-socialist wing (usually represented as the left) and the much more pragmatic, social democrat wing (usually a majority in the parliamentary party). That left–right conflict re-emerged as a major issue in the 1970s, with the feeling that the Wilson and Callaghan governments had failed the working class (Seyd, 1987). The tensions within the party that this created eventually led to a split, with a substantial portion of the social democratic wing defecting to the SDP. Within the Conservative Party, disillusion with the consensus and its links to what were seen as Britain's inexorable economic decline led to the creation of an ideological wing of what has always been very much a pragmatic party; that ideological wing captured the leadership of the party in 1975, and in the ensuing decade came to dominate the Cabinet and its entire policy-making process.

Creation of the SDP and then the Alliance in 1981 was seen as a counter to the tensions that the growing ideological wings were bringing to British politics – as exemplified in the development of the adversary politics model (Finer, 1975). The Alliance belief was that whereas the dominant elements in each party were pulling away from the consensus the electorate still remained very much middle-of-the-road. If a viable, centrist alternative could be presented to them, they would desert Conservative and Labour. They did, but not in sufficient numbers, and the momentum built up in 1983 was lost in 1987, in part through inept management.

The electoral geography

What of the electoral geography of the period? Its predominant feature was continuity; from 1922 through to 1979 the geography of each party's vote was strongly correlated with a single common pattern, as defined by principal components analysis (Johnston, 1983). That pattern was also closely correlated with the social and economic geography of the country, as Miller's (1977) detailed analyses have shown: the Labour Party was strongest in the more working-class areas and the Conservative Party in the more middle-class areas. But, as later analyses that combined survey with census and electoral data showed, the geography of voting was spatially more polarized than the geography of class: Labour tended to do even better than expected in the working-class areas (especially the coal- mining communities) while the Conservative strength was greater than anticipated in the middle-class and rural areas. Accounts of this polarization (such as Miller's, 1977) have tended to focus on the neighbourhood effect, but reconsiderations (see, for example, Johnston, 1986b) have argued for the importance of party as a mobilizing agent in communities, providing a context for both long-term socialization and short-term evaluation. In a strongly pro-Labour area, for example, where the party is well organized and trade unions are strong and active, and where the party has a majority on the local town and rural councils, people are more likely to be socialized into a pro-Labour set of attitudes; further, during any election campaign, the strength of pro-Labour sentiment will lead to an interpretation of events which is sympathetic to the party's views.

The creation of that geography was therefore a function of party activity (as Cox, 1970, demonstrated for Wales). The Labour Party mobilized the working class in the industrial constituencies, where the organizational structure of work provided a context for winning over large numbers of supporters, in close collaboration with the trade unions. It was less successful at winning such support in places where the working-class was less readily mobilized – in the rural areas and smaller towns, for example, and also in cities dominated by either service industries or small manufacturing plants – and where the general pro-Conservative milieux created by the middle-class, along with the greater prevalence of working-class deference, made for less commitment to the Labour cause. The result, then, was that compared to the national situation, Labour got more support from the working-class in the industrial constituencies than elsewhere. It also tended to get more support from other classes, in part because they responded to the local milieux and in part because many members of the middle class there were either upwardly mobile products of the local working class or had chosen to work and live there as a consequence of their political attitudes.

This geography was unchanging but not fixed: the general topography remained constant over the period, with the highs and lows of Labour support always in the same areas, but the amplitude of the surface altered as the relative fortunes of the parties waxed and waned. Thus at some general elections (notably 1945, 1950, 1951) Labour did much better at winning working-class support across the country than it did on average. This success can be linked to the process of retrospective voting. In 1945, as Thrift (1986) argues in detail, the electorate expressed its dissatisfaction with the economic conditions of the 1930s, and thereby was anti-Conservative; the new milieux advanced the Labour cause for many, especially those with relatively little earlier contact with it, and eased the process of shifting one's vote. The policies of Labour's government satisfied sufficient voters for it to retain much of its support in both 1950 and 1951, but the slight decline in support in some places and the vagaries of the British electoral system meant that it lost in 1951; it has never again received such support from the electorate.

Linked to the continuity in the geography of party support was the phenomenon of the uniform swing (Butler and Stokes, 1969), whereby between any pair of elections the shift in the percentage of the electorate supporting one of the two main parties was approximately the same in all constituencies. (Thus if at election I the percentages nationally were Conservative 54 and Labour 46 and at election II they were 49 and 51, this was represented as a 5 per cent swing against the Conservatives and towards Labour. That swing was uniform across the country.) This was interpreted by many as indicating a uniform political culture, that the voting trends were the same everywhere. In fact, as Butler and Stokes (1969) made clear, the uniform swing in percentage point terms meant that in constituencies where a party was strong it lost a smaller proportion of its share of the electorate than it did where it was relatively weak. (If there were a 5 per cent swing against the Conservatives, then in a constituency where the party had 70 per cent of the two-party vote at the first election it would have lost just over 7 per cent of that share – $(5/70) \times 100$ – whereas in one where it had only 30 per cent it would have lost nearly 17 per cent – $(5/30) \times 100$.) In relative terms, the uniform swing effect meant that parties did well where they were strong and badly where they were weak (Johnston, 1981; Johnston and Hay, 1982).

This 'variability within uniformity' has also been traditionally associated with the neighbourhood effect concept: where a party is strong, the process of 'conversion by conversation' (modelled by Johnston, 1976) will mean that it loses fewer votes (if the overall swing is against it) than it does where it is weak. But many dispute the operation of those processes (Dunleavy, 1979, is a particular critic), and seek other explanations. The geography of milieux provides a possible one: the stronger a party's support in an area, the more committed people are to it, and the less likely they are to desert it.

A third possible account links the geography of the uniform swing to retrospective voting. If we take it as axiomatic that a party in government will act in many of its policies in the interests of its main supporters, then presumably the people in the constituencies where it is strongest will be either the most satisfied – if the swing is towards the party – or least dissatisfied – if the swing is against it. Thus, if the belief that governments lose elections rather than oppositions winning them is valid, we can hypothesize that a government is most likely to lose an election in the constituencies where it is relatively weak – because those are the areas that will have benefited least from its policies.

What of the period 1979–87, when the polarization of the electorate has become much more marked (Johnston and Pattie, 1987; Johnston *et al.*, 1988)? In 1979, when Labour lost, the geography of retrospective voting postulated above clearly operated: in northern urban constituencies, in Scotland and in Wales the electorate was relatively satisfied with Labour's performance and voted accordingly; in the south-east (with the partial exception of Greater London) and the Midlands voters were much more dissatisfied and so voted for the Conservatives. Four years later, southerners were clearly much more satisfied than northerners with the performance of the Thatcher government, and much more optimistic about the future; they voted accordingly, extending the north–south divide. The same happened in 1987 (Johnston and Pattie, 1989).

The 1983 general election saw a new element in British electoral politics, however – the Alliance. Traditionally, a vote for the Liberal Party was a protest vote against the party in power by somebody unprepared to vote for the other main party (i.e. a dissatisfied Conservative supporter who felt unable to vote Labour: Himmelweit *et al.*, 1985). The authors of the 1983 British election study suggest that an Alliance vote then was not a protest vote, but a positive statement of support for its political ideology (Heath *et al.*, 1985). Further statistical analyses provide some support for this, but the proportion of Alliance voters who can be correctly predicted from knowledge of their attitudes is much less than is the case for Conservative and Labour (Johnston and Pattie, 1988). In 1983 and 1987 a vote for the Alliance was still a protest vote for many, it seems: in the south of England people dissatisfied with the economic situation and pessimistic about the future were much more likely to vote Alliance than Labour; farther north the reverse was the case. This pattern of who to vote for if dissatisfied widened the north–south divide.

The growing north–south divide of the 1980s is thus a continuation of an on-going trend (identified by Curtice and Steed, 1982; see also Curtice and Steed, 1988), accentuated by the growth of the Alliance. It is the product of the social and economic geography of the country that has resulted from the Tory policies. But there is a little more to it, because it is apparently linked to the process of dealignment, and the weakening of affinities with parties. Because of those weaker links people are much more ready to vote according to their evaluation of government policies – i.e. short-term evaluation – rather than long- term loyalties. Thus people in the prosperous south will vote Conservative, whatever their class, to reward the government for its success, with obvious consequences for Labour, which sees its electoral base there (already relatively weak) being further eroded.

Political consequences

This brings us finally to the consequences of those trends. The spatial polarization in voting patterns has produced a pattern of representation where the number of marginal seats has declined (Curtice and Steed, 1986; Johnston *et al.*, 1988). It would take swings of up to 10 percentage points away from the Conservative incumbents for the opposition parties to win sufficient seats so that the Tories no longer had a majority in the House of Commons. In many southern constituencies the Labour presence is now minimal, as measured by the number of deposits lost in the last two elections (a candidate's deposit is lost if he/she gets less than a fixed percentage, currently 5, of the votes cast), and the Alliance (on the basis of the 1987 results) does not pose a substantial threat to the Tory lead in many seats. So how do the opposition parties approach the next election?

One approach is to do relatively little that is innovatory and likely to attract either or both of internal dissension and wide public debates. The parties could sit back and wait either for an economic recession, in which case the Tories would lose substantial support according to the retrospective voting model, or a crisis of confidence in the Conservatives, either because of particular policies (on the health service, for example) or because of a decline in support for the leadership. But neither Labour nor the Alliance can be sure they would necessarily benefit from such a trend, in part because the electorate may either shift to the right or remain loyal to the Tories because of the perceived inadequacies of any alternative. The decade of Tory rule has not just resulted in a widening north–south divide in terms of economic and social well-being, it has also involved a major restructuring of social relations in the prosperous parts of the country, such that the foundations on which Labour could build are weak indeed, whereas the problems within the Alliance (replaced in 1988 by the merged Social and Liberal

Democrat Party but with a rump of the SDP remaining as an independent force) make its vote-winning potential difficult to assess.

There is widespread realization within the Labour Party that it must repackage itself if it is to present a viable challenge. But the nature of the package is hotly contested within the party, between those who see a need to promote socialism and those prepared to jettison socialism for a social democratic case based on morality, equity, and a more humane way of accommodating to capitalism. If either strategy were adopted it would involve problems. A socialist programme might sustain Labour's vote in the more depressed areas, but could it win votes where they are needed – in the affluent south? A social democratic programme might be more attractive to the relatively affluent, particularly if it could appeal to those who had recently voted for the Alliance, but would it lead to a split in the party and a loss of votes and seats in the north? Of course, parties have operated as coalitions of very different things to different areas, but it is doubtful whether that is feasible in a small, centralized country such as Great Britain.

Crewe (1986) has argued that Labour must choose its future direction with regard to both its political and its social strategy. With regard to the former, it must decide whether to stick to its traditional appeal (i.e. the politics of class conflict) or to develop a new one; its social strategy must either focus its attention on reviving working class conflict or appeal more widely to other classes. Combining the two provides four possible strategies (Table 8.1).

Table 8.1: Possible strategies for the British Labour Party

	Social Strategy	
Political Strategy	*Retrieve Working-class support*	*Appeal to other classes*
Traditional appeal	The class warfare model	The equality and justice model
New appeal	The populist authoritarian model	The liberatarian model The new technology model

1 A traditional appeal aimed at retrieving working-class support would mean reviving the politics of class warfare, seeking to win power via working-class solidarity behind an ideological appeal, as promoted by Benn in the 1980s leadership contests.

2 A new appeal but the same social strategy would downplay class conflict, and instead seek working-class solidarity on some other basis, such as the populist authoritarian model

propounded by Callaghan between 1976 and 1979.

3   Widening the traditional appeal to other classes would be based on other ideological grounds – such as equality and justice in the operation of market forces, as proposed by the Labour leadership after the 1987 defeat (see Mitchell, 1987).

4   Changing both social and political strategies, as with the 'new technology' programme outlined by Wilson in the 1970s and the libertarian model developed by Livingstone in the 1980s.

Crewe argues that neither of the first two is viable, because there is insufficient working-class support available to provide a parliamentary majority. Thus Labour must adopt a new social strategy, and possibly a new political strategy too.

Some parallels to this can be found in the new directions taken by the New Zealand Labour Party since 1984 (as described in Chapter 6), which has developed new political and social strategies. The parallels are not too close, however, because there the strategies were only introduced *after* Labour was elected to power in 1984 as a consequence of a decline in voter confidence in the incumbent National Party – a good example of retrospective voting leading to the election of a credible alternative. When elected, the New Zealand Labour Party had not developed new strategies: can the British Labour Party act similarly?

Crewe's advice is that it cannot. If Labour goes for a new social strategy it therefore seeks to break the current cleavage much more explicitly than in the past. By developing policies that make it acceptable to a wider spectrum of society than heretofore (including the affluent working class in southern England) it could win substantial numbers of votes when the Conservative Party suffers a loss of confidence and the disadvantages of the retrospective voting model. But in doing so, Labour will no longer have the firm social base of earlier decades: the order provided by the class cleavage will have been eroded, and the resulting potential volatility will make the pattern of voting much more the whim of voter reaction to recent policies; to the extent that those policies are spatially uneven in their impact the electoral geography may become more volatile too.

To some, the only answer – if the main goal is to defeat the Conservatives – is some kind of Labour–Alliance pact, which will build on the two parties' relative strengths in constituencies where victory for an 'anti-Conservative alliance' looks feasible. For Labour, however, this would not only mean working with people who had deserted the party less than a decade ago, but it would also mean admitting that victory alone is not possible. While many Labour members may be prepared to be more pragmatic than ideological in order to win, would they be that pragmatic?

Of course, they will not be fighting in a vacuum. Since re-election in 1987 the Conservative Party has carried forward policies aimed not just at extending prosperity to the north but also at continuing the process of social restructuring – in, for example, the erosion of the powers of trade unions and local government. In this they are eroding Labour's base in its heartland, eating into the Labour side of the class cleavage. The elections of 1983 and 1987 pushed Labour into a spatial ghetto, where it sustained majority support from 'its class'; by 1991–2 that spatial ghetto may be restricted in size, further constraining Labour to minority status. How then might Labour react?

Return to Lipset and Rokkan

I began this chapter with a reference to the problems of the Alliance as it sought to 'break the mould' of British politics in the 1980s. This led to a discussion of what the mould is, for which Lipset and Rokkan's concept of electoral cleavages provided a framework which has been widely employed in much recent writing about electoral behaviour.

A review of Lipset and Rokkan's model, with particular reference to single-member constituency, plurality electoral systems in which the dominant parties are pragmatic rather than ideological, identified a number of problems. Cleavages are necessary, according to the model, because they structure political conflict within society and are thus a source of social order. If the cleavages are too well defined, however, that order is counter productive, since the minority group will be unable to gain access to power. There must be sufficient flexibility across the divide so that both the dominant parties in such systems can expect electoral victory.

The source of that flexibility was identified as the concept of retrospective voting, whereby people evaluate parties at general elections in terms of past performance relative to the voters' own interests. Since most parties when in power will produce policies that benefit their core supporters, this retrospective voting can account for the relative uniformity of swing observed in Britain for several decades. Recently that pattern has broken down, however, and has led to the partial dissolution of the dominant cleavage and with it a major change to the electoral geography. The two parties have become more ideological. The successful one in the 1980s has succeeded by gaining many votes from all occupational classes in the parts of the country where its ideologically based policies have brought about prosperity for a majority; the main opposition party – Labour – has increasingly got support only in those areas of relative deprivation. Over the 1980s, therefore, class heterogeneity in voting behaviour has increased very substantially.

A major consequence of this shift has been the change in the number of marginal seats, such that it is now very difficult to see how Labour could win the next election – either alone or in some pact with the Alliance, which has been the source of relatively dissatisfied voters' ballots in the prosperous part of Britain. The options open to Labour are difficult. Analysis of recent trends in New Zealand was undertaken to see whether they offered any pointers to possible Labour strategies in Britain. It showed another country with increased class heterogeneity in voting behaviour (in New Zealand the middle class has become more heterogeneous; in Britain, it is the working class that has done so) as a result of the Labour Party's package of economic policies to win middle class support, defence and cultural policies to appease its own left wing, and promised social policies to retain the (grudging at present) support of its traditional working class supporters. There is one major difference between the two, however: the New Zealand Labour Party's substantial pragmatic shift in policy has taken place when it is in power, so that the electorate voting retrospectively can evaluate what it has done, not what it might do. To this extent, it is the same as the British Conservative Party: both may lose when the evaluations are negative, but the opposition parties cannot sit back and wait for that to happen, they must try to hasten it.

The basic problem with the cleavage model is that it does not contain a clear mechanism for change. Lipset and Rokkan, as noted above, argued that cleavages had been stable for 40 years or more in several Western liberal democracies. They did not contend that this 'freezing of political alternatives' (p. 54) was in any way permanent, and recognized that there 'will clearly be greater fluctuations than before' (p. 56) but said very little about the nature – cause and resolution – of those fluctuations. The American experience shows that major realignments can take place to unfreeze a situation where one group is permanently in the minority, but there the cleavages have never been as deep as in Europe (Archer and Taylor, 1981). All this chapter has been able to do is explore the nature of realigning processes in Great Britain, as a beginning of the process of incorporating a clearer dynamic to the Lipset and Rokkan model.

References

Alford, R. R. (1963) *Party and Society*. Chicago: Rand McNally.
Archer, J. C. and Taylor, P. J. (1981) *Section and Party*. Chichester: Wiley.
Bogdanor, V. and Butler, D. (eds.) (1983) *Democracy and Elections*. Cambridge: Cambridge University Press.
Bradley, I. (1981) *Breaking the Mould?* Oxford: Martin Robertson.
Butler, D. and Stokes, D. (1969) *Political Change in Britain: Forces*

changing Electoral Choice first edition, London: Macmillan.

—— (1974) *Political Change in Britain*. second edition, London: Macmillan.

Cox, K. R. (1969) The voting decision in a spatial context. In C. Board *et al.*, editors, *Progress in Geography, Volume 1*, London: Edward Arnold, 81–117.

—— (1970) Geography, social contexts and voting behavior in Wales, 1861–1951. in E. Allardt and S. Rokkan (eds) *Mass Politics*, New York: The Free Press, 117–159.

Crewe, I. (1986) 'On the death and resurrection of class voting: some comments on "How Britain votes"', Political Studies 34: 620–38.

—— and Denver, D. (eds.) (1985) *Electoral Change in Western Democracies*. London: Croom Helm.

Curtice, J. (1988) Political partisanship. in R. Jowell, S. Witherspoon and L. Brook, (eds.) *British Social Attitudes: the 1986 Report*. Aldershot: Gower 38–53.

—— and Steed, M. (1982) Electoral choice and the production of government. *British Journal of Political Science* 12: 249–98.

—— (1986) Proportionality and exaggeration in the British electoral system. *Electoral Studies*, 5, 209–28.

—— (1988) The results analysed, in D. Butler and D. Kavanagh (eds.) *The British General Election of 1987*. London: Macmillan.

Dalton, R., Flanagan, S.C. and Beck, P. (eds.) (1984) *Electoral Change in Advanced Industrial Societies*. Princeton: Princeton University Press.

Downs, A. (1957) *An Economic Theory of Democracy*. New York: Harper & Row.

Dunleavy, P. (1979) The urban basis of political alignment. *British Journal of Political Science* 9: 409–43.

Finer, S. E. (ed) (1975) *Adversary Politics and Electoral Reform*. London: Anthony Wigram.

Fiorina, M. P. (1981) *Retrospective Voting in American National Elections*. New Haven, Conn.: Yale University Press.

Gudgin, G. and Taylor, P. J. (1979) *Seats, Votes and the Spatial Organization of Elections*. London: Pion.

Harrop, M. and Miller, W. L. (1987) *Elections and Voters*. London: Macmillan.

Heath, A., Jowell, R., and Curtice, J. (1985) *How Britain Votes*. Oxford: Pergamon.

Himmelweit, H. T. Humphreys, P., and Jaeger, M. (1985) *How Voters Decide*. Milton Keynes: The Open University Press.

Inglehart, R. (1977) *The Silent Revolution*. Princeton, N.J.: Princeton University Press.

—— (1981) 'Post-materialism in an environment of insecurity'. *American Political Science Review* 75: 880–99.

Johnston, R. J. (1976) 'Contagion in neighbourhoods'. *Environment and Planning* A 8: 581–6.

—— (1979) *Political, Electoral and Spatial Systems*. Oxford: Oxford University Press.

—— (1981) 'Regional variations in British voting trends 1966–1979'. *Regional Studies* 15: 25–32.

—— (1982) 'Political geography and political power'. In M. J. Holler (ed.) *Power, Voting and Voting Power*. Vienna: Physica-Verlag, 289–306.

—— (1983) 'Spatial continuity and individual variability'. *Electoral Studies* 2: 53–68.

—— (1985) *The Geography of English Politics*. London: Croom Helm.

—— (1986a) 'A space for place (or a place for space) in British psephology'. *Environment and Planning A* 19: 599–618.

—— (1986b) 'The neighbourhood effect revisited'. *Environment and Planning D: Society and Space* 4: 41-55.

—— (1987a) *Money and Votes*. London: Croom Helm.

—— (1987b) 'Dealignment, volatility and electoral geography'. *Studies in Comparative International Development* 22: 3–25.

—— and Hay, A.M. (1982) 'On the parameters of uniform swing in single-member constituency electoral systems'. *Environment and Planning A* 14: 61–74.

—— and Honey, R. (1988) 'The 1987 general election in New Zealand and the demise of electoral cleavages', *Political Geography Quarterly* 7: 363–8.

—— O'Neill, A. B. and Taylor, P. J. (1987) 'The geography of party support: comparative studies in electoral stability', in M. J. Holler, (ed.) *The Logic of Multiparty Systems*, Dordrecht: Kluwer, 265–80.

—— and Pattie, C. J. (1987) 'A dividing nation? An initial exploration of the changing electoral geography of Great Britain, 1979-1987'. *Environment and Planning A* 19: 1000-13.

—— (1988) 'Are we all Alliance now? Discriminating by discriminant analysis'. *Electoral Studies*, 7: 27–32.

—— (1989) 'A nation dividing? Economic well-being, voter response, and the changing electoral geography of Great Britain'. *Parliamentary Affairs* 41: 37–57.

—— Pattie, C. J. and Allsopp, J. G. (1988) *A Nation Dividing? The Electoral Map of Great Britain, 1979–1987*, London: Longman.

Key, V. O. (1955) 'A theory of critical elections'. *Journal of Politics* 17: 3–18.

—— (1966) *The Responsible Electorate*. New York: Vintage.

Kirby, A. M. and Taylor, P. J. (1976) 'A geographical analysis of the voting pattern in the EEC referendum. 5 June, 1975'. *Regional Studies* 10: 183–92.

Lipset, S. M. (1970) 'Political cleavages in "developed" and "emerging" polities', in E. Allardt and S. Rokkan (eds.) *Mass Politics*, New York: Free Press, 23–44.

—— (1984) *Political Man*. London: Heinemann.

—— and Rokkan, S. (1967) 'Cleavage structures, *Party systems and voter alignments*', in S. M. Lipset and S. Rokkan, (eds.) Party Systems and Voter Alignments: Cross-national Perspectives, New York: Free Press, 3–64.

McAllister, I. and Rose, R. (1984) *The Nationwide Competition for Votes: the 1983 British Election*. London: Pinter.

Miller, W. L. (1977) *Electoral Dynamics*. London: Macmillan.

—— (1984) 'There was no alternative'. *Parliamentary Affairs* 32: 376–82.

Mitchell, A. (1987) 'Beyond socialism'. *Political Quarterly* 58: 389–403.

Osei-Kwame, P. and Taylor, R. J. (1984) 'A politics of failure: the political geography of Ghanaian elections, 1954-1979'. *Annals of the Association of American Geographers* 74: 574–89.

Parsons, T. (1959) 'Goal theory in sociology', in R.K. Merton (ed.) *Sociology Today, New York: Basic Books*, 39–78.

Rae, D. W. (1971) *The Political Consequences of Electoral Laws*. New Haven: Yale University Press.

Pulzer, P. (1967) *Political Representation and Elections in Britain*. London: Allen & Unwin.

Rae, D. H. (1971) *The Political Consequences of Electoral Laws*. New Haven, Conn.: Yale University Press.

Rose, R. and Urwin, D. (1975) *Regional Differentiation and Political Unity in Western Nations*. Beverly Hills, Cal.: Sage Publications.

Sanders, D., Ward, H. and Marsh, D. (1987) 'Government popularity and the Falklands war: a reassessment'. *British Journal of Political Science* 17: 281–313.

Seyd, P. (1987) *The Rise and Decline of the Labour Left*. London: Macmillan.

Taylor, P. J. (1986) 'An exploration into world-systems analysis of political parties'. *Political Geography Quarterly*, Supplement to 5, S5–S20.

—— and Johnston, R. J. (1979) *Geography of Elections*. Harmondsworth: Penguin.

Thrift, N. J. (1986) 'Little games and big stories: accounting for the practice of personality and politics in the 1945 general election', in K. Hoggart and E. Kofman, (eds.) *Politics, Geography and Social Stratification*. London: Croom Helm, 56–143.

Chapter nine

The cleavage model and electoral geography: a review

A. Lijphart

That the cleavage model, first stated more than two decades ago by Seymour Martin Lipset and Stein Rokkan (1967) has had a strong and enduring impact on the social sciences is once again demonstrated by the chapters in this section of the book. Probably a majority of scholars have tended to accept the major elements of the model and have cited it approvingly, but it has also encountered strong criticism and even rejection. One of the principal detractors is Michal Shamir (1984: 35), who argues against the famous Lipset–Rokkan thesis of the 'freezing' of party systems since the 1920s and reaches exactly the opposite conclusion: 'party systems ... have never really been frozen'. In my opinion, however, Shamir attacks an overly broad interpretation of the 'freezing' proposition instead of the proposition that Lipset and Rokkan intended.

Shamir's criticisms deserve a closer look, and I shall return to this matter below. First, I shall try to define the cleavage model as clearly as possible and to formulate the major propositions that it entails. In doing so I shall also state my own criticisms of certain aspects of the model, which are by and large very mild criticisms, concerning errors of omission rather than of commission. Three separate, although obviously related, propositions can be distinguished. The first concerns the four issue dimensions that Lipset and Rokkan claim to have been predominant in Western party systems. The second and third propositions both have to do with the freezing of party systems: the party organizations established in the 1920s have been 'frozen' since then, and the issue alternatives represented by the party systems have similarly remained basically the same.

Issue dimensions in party systems

The first proposition is a miracle of elegance: two successive revolutions – the national and industrial revolutions – have each generated two dimensions of oppositions, yielding a total of four of these basic dimensions and eight basic alternatives on the dimensions: central v. peripheral culture, nation-state vs. church, industrial v. agrarian interests, and workers v. employers. Somewhat more simply formulated, the corresponding issue dimensions are the cultural–ethnic, religious, urban–rural, and socio-economic dimensions. Lipset and Rokkan's thesis is highly persuasive as far as both the undoubted importance of the four basic issue dimensions and their developmental origins are concerned. My only criticism is that these four do not exhaust the issue alternatives characterizing Western party systems and that four additional dimensions can be identified – happily maintaining the numerological elegance of the analysis! I hasten to add that Lipset and Rokkan do not claim that their four dimensions are the *only* dimensions, but they repeatedly refer to them as the 'decisive dimensions of opposition' (1967: 33, 34), and, with one minor exception, fail to mention any other dimensions.

The exception is that Lipset and Rokkan (1967: 54) observe considerable disillusionment during the 1960s and a potentially new dimension of opposition fed, in particular, by 'the widespread disagreements with the national powers-that-be over foreign and military policy'. Obviously, however, foreign policy issues have divided political parties in many countries for a much longer time. The most prominent example is the Irish case discussed by O'Loughlin and Parker. Here the controversy over the treaty establishing national independence was essentially a foreign policy dispute: divergent attitudes toward the status of Northern Ireland and the republic's relationship with the British government. This controversy created the two major Irish parties in the 1920s, Fianna Fail and Fine Gael, which have remained Ireland's dominant political forces.

In addition to neglecting the interaction of domestic politics and foreign policy, Lipset and Rokkan ignore the importance of anti-democratic parties. The Weimar Republic offers the clearest example of very large parties of this kind, but elsewhere, too, anti-democratic parties on the extreme right and left have played a significant role. This entails a sixth issue dimension: that of regime support.

Two other dimensions are suggested in Johnston's chapter: the 'growth of the state' and 'post-industrialism'. I am not persuaded that the former should be regarded as a separate dimension; it can be accommodated under the traditional left–right socio-economic dimension with the left representing the pro-growth alternative and the right the anti-growth position. On the other hand, post-industrialism is not only a new and clearly different force but also contains – contrary to my own earlier interpretation (Lijphart, 1984: 139–41) – two separate issue dimensions: participatory democracy and environmentalism. The new 'left-libertarian' parties (to use Herbert Kitschelt's, 1988, label), most notably the West German Greens, frequently advocate radical policies with regard to both dimensions, but the two dimensions are logically and conceptually quite distinct.

It seems to me that Passchier and van der Wusten adhere too closely to the limited four-dimensional Lipset–Rokkan model when, for the purposes of their analysis, they count the new Dutch party Democrats '66 (D'66, founded in 1966 and represented in parliament since 1967) as one of the Liberal parties. Their classification of the parties into the four groupings of Socialists, Liberals, Catholics, and Protestants (leaving aside the residual 'Uncommitted' category) is based on the socio-economic and religious dimensions – and these have indeed been the most salient issue dimensions in Dutch party politics since the 19th century. Even if we accept this two-dimensional framework, the combination of Liberals and Democrats '66 appears questionable: on the socio-economic dimension, D'66 has tended to take a left-of-centre instead of a right-of-centre position and has therefore been closer to the Socialists than to the Liberals. Since its first electoral success in 1967, D'66 has participated in both of the centre-left Cabinets in which the Socialists also took part (1973–7 and 1981–2) but never joined any of the centre-right Cabinets that included the Liberals (1967–73, 1977–81, and from 1982 on).

But a more fundamental objection to the classification of D'66 as a Liberal party is that this new party simply does not fit the traditional issue dimensions of Dutch politics very comfortably. Its principal *raison d'être* has been democratic renewal, and we need the additional participatory democracy dimension in order to understand and classify it properly. In terms of the two old dimensions it is clearly a non-religious party but on socio-economic issues it is primarily a centre party rather than either a rightist or leftist party. Instead of explaining its alliances with the Socialists in terms of its centre-left position on the socio-economic dimension, it is more convincing to stress that the Socialists – unlike the Liberals – have adopted a position on the progressive or 'renewal' side of the participatory-democracy dimension. Hence it is on this new dimension that D'66 and Socialists have been natural partners.

To sum up the argument so far: the Lipset–Rokkan thesis concerning the four decisive issue dimensions dividing Western political parties has to be amended by adding four dimensions – the foreign policy, regime support, participatory democracy, and ecological dimensions. This does not mean that, especially from the long-term perspective of the entire period from the 1920s to the end of the 1980s, the two sets of dimensions are of equal importance. The two post-materialist dimensions are of only recent origin, of course, and the foreign policy and regime support dimensions have become less salient since the 1960s. On the other hand, I have found that in the party systems of 21 advanced industrialized democracies (the Western democracies plus Japan) in the 1945–80 period, partisan differences occurred more frequently on the foreign policy dimension than on the Lipset–Rokkan cultural–ethnic dimension and as frequently as on Lipset and Rokkan's urban–rural dimension (Lijphart, 1984: 139).

The freezing of Western party systems

Let us now turn to what Lipset and Rokkan call the 'freezing' of Western party systems. This is often regarded as a single proposition, but in fact two separate theses must be distinguished. Lipset and Rokkan (1967: 50, italics omitted) first state that 'the party systems of the 1960s reflect, with few but significant exceptions, the cleavage structures of the 1920s', but they proceed to distinguish between (1) 'the party alternatives' which 'are older than the majorities of the national electorates' and (2) 'in remarkably many cases the party organizations' which similarly originated in the 1920s. The first of these is essentially an extension of the proposition concerning the four decisive issue dimensions. What Lipset and Rokkan assert is that not only were these four dimensions crucial in the formation of Western party systems but also that, once those party systems were in place, the four dimensions remained the principal differentiators among the parties. In my earlier discussion of the four-dimensional Lipset–Rokkan framework, I have already commented on both these aspects, and my four suggested additional dimensions divide equally between two older dimensions, dating back to the 1920s (foreign policy and regime support) and two newer dimension originating in the 1960s and 1970s (participatory democracy and environmentalism). The only further point that should be made here is that Lipset and Rokkan, writing in the 1960s, should in all fairness not be blamed for not discerning or foreseeing the development of these new dimensions.

As far as the freezing of the party *organizations* is concerned, it is worth stressing that Lipset and Rokkan themselves admit that there are exceptions. They state that, as already cited in the previous paragraph, the party organizations have been frozen since the 1920s 'in remarkably many cases'. This means that the Lipset–Rokkan thesis cannot be invalidated by the presentation of one or a few counter-examples.

Another inadmissible line of attack has been to define the freezing theory in much too specific terms – terms that Lipset and Rokkan never stated or intended – and then to criticize the redefined version of the theory. Shamir (1984: 39–42) claims that Western party systems were never frozen, since they can be shown to have undergone considerable fluctuation with regard to levels of voter support, fragmentation, and ideological polarization. However, what Lipset and Rokkan have in mind is that the significant parties have remained significant parties – not that they have not been subject to ups and downs in popular appeal. Hence evidence of unstable voter support does not invalidate the freezing proposition. Similarly, one cause of changing party system fragmentation is that parties fluctuate in size. As Shamir measures it, a three-party system with three parties of equal size is more fragmented than a three-party system with a 50–25–25 per cent pattern of strength. As long as the three parties in the system are basically the same parties, such a change in fragmentation does not contradict the Lipset–Rokkan thesis either. Finally, evidence of changing ideological polarization cannot invalidate the thesis. Lipset and Rokkan state that the party alternatives have been frozen, not that on these alternatives the parties have always maintained the same distance. To his credit, Shamir (1984: 35) admits that the theory he attacks is really not the one advanced by Lipset and Rokkan, but the 'freezing assumption ... adapted and expanded by others implicitly or explicitly'.

Not only do Lipset and Rokkan (1967: 51–53) allow exceptions to the freezing of party organizations, but they also identify these exceptions as occurring in the major countries – France, Germany, Italy, and Spain – and they call them 'spectacular'. Furthermore, they subject these deviant cases to analysis in order to try to explain why the exceptions occur. They find that the crucial explanatory factor is the establishment of mass organizations which usually took place in the 1920s. Where, for one reason or another, strong mass organizations were not formed – as on the right of the political spectrum in France and Germany – the parties were not frozen and 'a great deal of leeway for innovation in the party system' continued to exist. This is in line with Joseph Schumpeter's (1955: 64–5, 98) well known proposition – illustrating 'the ancient truth that the dead always rule the living' – that political and social institutions, 'once firmly established, tend to maintain themselves and to continue in effect long after they have lost their meaning and their life-preserving function'. In other words, for Lipset and Rokkan the decisive factor for the freezing of party organizations is not the origin of the organizations in the 1920s but the establishment of strong mass organizations.

According to this improved version of the freezing theory, cases like the French Gaullists and the German CDU/CSU, both originating in the 1940s, are no longer deviant. With regard to the latter, Lipset and Rokkan (1967: 52) express some doubt. They state that 'with the establishment of the regionally divided CDU/CSU the Germans were for the first time able to approximate a broad conservative party of the British type', and that this party proved to be highly effective 'at least until the debacle of 1966'. With the advantage of hindsight, it is clear that 1966 was merely a temporary setback and that the late arriving but firmly established CDU/CSU fits the frozen pattern without any difficulty.

A second explanation advanced by Lipset and Rokkan (1967: 30–3, 55) has to do with the electoral system – also highlighted in Johnston's chapter – which can affect the *degree* to which party systems tend to remain frozen. Plurality and majority systems entail high barriers to new parties since these almost always begin as small parties; regionally concentrated parties are an exception but they are likely to remain small and peripheral. Hence these high-barrier systems protect the existing frozen pattern. On the other hand, systems of proportional representation offer much better opportunities for challenges to the existing parties. When we look at the evidence of the countries with proportional representation, two conclusions stand out. First, it is a tribute to the strength of the freezing theory that even in these low-barrier systems the existing parties tend to maintain themselves very well once they are firmly established and organized. Second, it is also significant that the new parties representing the post-materialist positions have been at least moderately successful only in the low-barrier systems. Kitschelt (1988: 200) finds 'clear cases' of successful left-libertarian parties in eight countries: Austria, Belgium, Denmark, Germany, Iceland, Luxembourg, the Netherlands, and Norway. All eight use proportional electoral systems.

The cases of Ireland and New Zealand

Two of the chapters in this section can be read as extended tests of the freezing proposition: the chapter on Ireland by O'Loughlin and Parker and the chapter on New Zealand by Honey and Barnett. To what extent do these two cases offer support for – or to what extent do they weaken – the Lipset–Rokkan theory?

I regard the evidence of the Irish case as a striking confirmation of the freezing theory – not just as giving it 'considerable support', as the authors modestly state – with one minor exception. As indicated earlier, the split between Fianna Fail and Fine Gael on the treaty was essentially a division on the foreign policy dimension, which has to be added to the basic four Lipset–Rokkan dimensions. Otherwise the freezing explanation fits the Irish facts beautifully in three respects. First, the party system that emerged in the 1920s can be explained in terms of the relative salience of the four dimensions plus the foreign policy dimension. Three of the dimensions were irrelevant because all of Ireland was peripheral, rural, and church-oriented, and the socio-economic dimension was overshadowed first by national unity and subsequently by the split on the treaty. Second, the analysis of the referenda shows that the four Lipset–Rokkan dimensions did emerge after the 1920s as issue dimensions of some basic importance in Irish politics. But, third, these new dimensions did not affect the party system – which remained frozen not because of the absence of important alternative issue dimensions but in spite of the existence of those dimensions.

The evidence of the New Zealand case is more ambiguous. The swing to the right by the Labour Party is certainly a remarkable, and may even be a unique, phenomenon. The only parallel that comes to mind is the German Free Democratic Party, which made a sharply leftward move in the late 1960s followed by a swing to the right in the early 1980s. Both of these moves were quite risky, and the FDP experienced great difficulty in maintaining its voting support. Two differences between the FDP and the New Zealand Labour Party stand out. First, the FDP's shifts were relatively small, entailing moves from slightly right of centre to slightly left of centre and the other way round; in New Zealand the Labour Party jumped over the National Party and moved all the way from left to right. Second, in contrast with the electoral problems faced by the FDP, the Labour Party scored a spectacular victory. One explanation of the contrast must be the electoral system. The protection that the plurality system offers to the two large parties is so strong that they can afford to make sudden and major shifts.

Do the actions and the success of the New Zealand Labour Party contradict the freezing theory in any way? My argument is that there are no contradictions and that, in one sense, the evidence can even be read as a strong confirmation of the theory. The Labour Party's move from left to right was highly unusual, but it occurred in the same socio-economic dimension that had been dominant in New Zealand party politics for a long time. Since Lipset and Rokkan argue that the dimensions themselves tend to remain frozen but not that the parties' positions on those dimensions will always be frozen, the actions of the Labour Party do not violate the Lipset–Rokkan theory – although it would be exaggerated to read them as confirmation. However, the fact that, in spite of Labour's big ideological shift, the established two-party system was not only maintained but even strengthened in the 1987 election does does constitute strong confirming evidence.

Conclusions

I agree with Johnston's contention that the Lipset–Rokkan model is not dynamic and that it does not contain a 'mechanism for change'. But its static nature has also been its strength. It predicts things to remain static and frozen – a prediction that by and large has been borne out. To make the model more dynamic would require explaining the exceptions to the model – but, with few amendments that I have offered, there are simply not many exceptions left and hence there is remarkably little of a dynamic nature that requires explanation.

References

Kitschelt, H. P. (1988) 'Left-libertarian parties: explaining innovation in competitive party systems', *World Politics* 40, 2: 194–234.

Lijphart, A. (1984) *Democracies: Patterns of Majoritarian and Consensus Government in Twenty-one Countries*. New Haven, Conn.: Yale University Press.

Lipset, S. M. and Rokkan, S. (1967) 'Cleavage structures, party systems, and voter alignments', in S. M. Lipset and S. Rokkan, (eds.) *Party Systems and Voter Alignments: Cross-national Perspectives*. New York: Free Press, 1–64.

Schumpeter, J. (1955) *Imperialism and Social Classes: two Essays*. trans. H. Norden. New York: Meridian Books.

Shamir, M. (1984) 'Are Western party systems 'frozen'? A comparative dynamic analysis', *Comparative Political Studies* 17, 1: 35–79.

Part III

American exceptionalism

Chapter ten

Populism and agrarian ideology: the 1982 Nebraska corporate farming referendum

Rebecca S. Roberts, Frances M. Ufkes, and Fred M. Shelley

Throughout American history, farmers have turned to political activity to cope with the vagaries of the market. Farmer activism heightens during periods of particular economic stress in the agricultural sector. Agrarian political activism in the United States has generally been directed against the increasing concentration of wealth and economic power in the hands of the rich at the expense of small family farmers. Farm activists have been less critical of the institution of private property and the process of capital accumulation itself; thus American farm protest is anti-monopoly yet not anti-accumulation.

The anti-monopoly attitude of many farmers has been expressed in strong opposition to the ownership of farmland by corporations. Central to American agrarian ideology is the notion that individual freeholders constitute the backbone of the economy. Many farmer activists view *corporate* farm ownership as incompatible with agrarian ideals. Agrarian protesters have often linked opposition to corporate farm intrusion with calls to save the family farm to preserve an essential aspect of the American heritage.

In 1982 the voters of Nebraska enacted Initiative 300, known popularly as the Nebraska Family Farm Amendment, which restricted the ownership of farmland in that state by corporate interests. Nebraska joined eight other Mid Western and Great Plains states that had enacted either constitutional or statutory restrictions on non-family corporate farmland ownership. However, support for the initiative varied substantially within the state (Fig. 10.1). It was strongest in the rural areas of eastern and central Nebraska – places that have been particularly subject to economic stress in recent years. The purpose of this chapter is to examine the geographical distribution of votes for and against Initiative 300, and in doing so to apply a social-theoretic perspective to the position of Mid Western producers within the changing political economy of the United States.

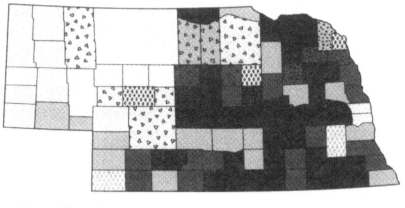

Percent Favouring

☐	0.0 to 49.9	▓	Traditional Corn Belt Region
▨	50.0 to 60.0	▓	Great Plains Region
■	60.1 to 66.0	▓	Deleted from Regression Analysis
■	Greater than 66.0		

Figure 10.1: Voting on Initiative 300, by county
Source: Nebraska Blue Book

Social theory and agriculture in the Mid West

Political activism among American farmers is closely associated with feelings of exploitation, powerlessness, and alienation. Farmers have tended to attribute their financial hardships to distortions in the market place and to the unchecked power of corporations and monopolies (Goodwyn, 1986). Agrarian political culture views the state as an inherently neutral force whose roles are arbitrating between social classes and stabilizing market imperfections. Hence agrarian discontent tends to be channelled through the existing political system.

Gramsci (1971) examined the power of consciousness and ideology to facilitate the domination of capitalist social relations. He argued that coercion and oppression are expressed in a number of forms, including the 'level of ideas'. Ideas and belief systems can become forces of oppression and are especially ominous when ideologies become hegemonic and garner 'the consent of the dominated'. In such cases, ideology influences the ways in which discontent with existing social relations is channelled. Thus agrarian social struggle, like other social movements, is best examined in the context of the political economy of capitalism. This necessitates an investigation of the social relations of agriculture, the position of agriculture in capitalist development, and the application of corresponding social theory to understanding social movements that initiate changes in public policy.

The philosophical underpinnings of American agrarian ideology can be found in the writings of Locke, Hobbes and Bentham and in works from the French physiocratic tradition (Wood, 1984). Thomas Jefferson and other Democratic Republicans believed that certain enduring values were associated with the agrarian way of life. For the Jeffersonians, a nation of self-sufficient small farmers was one in which democracy would thrive. The ideal society would be composed of resourceful, self-reliant, yet community-oriented people:

> The codification of American agrarianism by Jefferson and his contemporaries hinged upon several crucial elements: economic independence of the family farmer in the form of private property, an inextricable bond between the family farm and democracy, and the ideal of the simple, egalitarian community
>
> (Buttel and Flinn, 1976: 474).

Agrarian values were strongly imbued with the virtues of property ownership. By toiling on his own land the yeoman farmer produces something uniquely his, which civil society has a duty to protect. Such notions and values contribute to what MacPherson (1962) has called 'possessive individualism'. Linkages between property ownership and *community* values contribute to the importance of local political activity in a small, democratic community. The importance of the family farm to the survival of democracy remains evident in American myths, values, and symbols (Bellah *et al.*, 1985; Craig and Phillips, 1983; Goldman and Dickens, 1983).

Although most agricultural commodities continue to be produced by family-owned farms, the material reality of the family farmer can differ substantially from such myths and images. The Mid Western farmer has a contradictory class position (Fig. 10.2). By owning land and implements the farmer represents capital; by working on the land the farmer is labour. In effect, farmers are owners of small businesses competing with one another in a market over which they have no control. Competition leads to overproduction, which results in declining commodity prices. In order to compensate for price reductions, farmers invest in new technologies, purchase or rent additional land, and otherwise attempt to increase production levels or cut production costs. The accrued benefits of such actions are short-lived, however, as these efficiencies stimulate additional production increases and price reductions, placing farmers on a 'technological treadmill'.

Capitalist development within agriculture leads to differentiation of the social relations of production (Fig. 10.2). The complexity of this process was addressed by Marx (1982a, b), Lenin (1977), and Kautsky

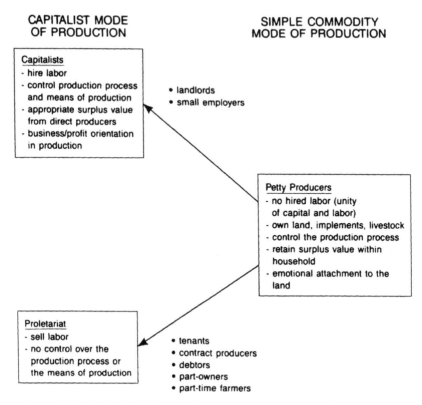

CAPITALIST MODE
OF PRODUCTION

SIMPLE COMMODITY
MODE OF PRODUCTION

Capitalists
- hire labor
- control production process
  and means of production
- appropriate surplus value
  from direct producers
- business/profit orientation
  in production

• landlords
• small employers

Petty Producers
- no hired labor (unity
  of capital and labor)
- own land, implements, livestock
- control the production process
- retain surplus value within
  household
- emotional attachment to the
  land

Proletariat
- sell labor
- no control over the
  production process or
  the means of production

• tenants
• contract producers
• debtors
• part-owners
• part-time farmers

Figure 10.2: Capitalist differentiation in the class structure of
agriculture

Source: Adapted from Mooney (1983)

(1980) and has been applied to the Mid West and Great Plains by Mann
and Dickinson (1978), Mooney (1978, 1982, 1983), and Goss *et al.*
(1980). At issue is the persistence of petty production relations in the
Mid West and Plains states, the possible routes that further
differentiation will take within agriculture, and the progressive potential
of producers in various, often contradictory, class locations.

Expansion of production on the part of some farmers forces the less
efficient ones out of agriculture. This leads to the gradual consolidation
of agricultural production on larger farms and a corresponding reduction
in the number of farmers. Some producers will expand, acquire more
land or increase their herds, and become small employers (Mooney,
1983). A few may become farm managers for other landowners, hiring
labour to carry out agricultural operations, and replacing an emotional
attachment to the land with a corporate business orientation. Some
producers may lose their land or may take off-farm employment in

addition to maintaining agricultural operations. Owner-operators who go heavily into debt in order to expand (e.g. to finance land acquisition or to buy additional production assets such as confinement sheds and irrigation equipment) also move along this route of dispossession (Mooney, 1983, 1986). Dispossession can be manifested directly as a loss of the means of production (e.g. through farm foreclosure) or through the loss of autonomy and control of the production process (e.g. through contract farming). The result of production expansion and dispossession is an increased commercialization of agriculture that stands in sharp contrast to the Jeffersonian ideal of the self-sufficient agrarian community.

Populism, farmer activism, and the American political economy

Agrarian political activity has characterized the rural sector throughout the history of the United States. During the early years of independence, however, much agrarian protest was deflected by the availability of cheap land along the western frontier. Although farmers complained of abuse from bankers, railroads, lawyers, and the government in general, '... those earlier days were the days of cheap lands, and when things went wrong the disgruntled could seek solace in a move to the West' (Hicks, 1931: 95). The closing of the frontier in the late 1800s brought an end to the availability of cheap land, thus removing the option of pulling up stakes and moving on. It was at this time that agrarian political protest reached its height. In short order, the Farmers' Alliance, the Grange movement, the Greenback movement, the Populist Party, and other social movements enlisted substantial support in rural farming areas in the United States.

Such movements, often referred to by the generic term 'populism', have focused on specific characteristics of capitalism rather than its underlying economic foundations. Ideologically, these are 'reformist' efforts rather than counter-hegemonic movements in the Gramscian sense. Populist activities have repeatedly been directed against the concentration of economic and political power. Local targets of agrarian discontent in the Mid West and Great Plains regions have included grain milling and food processing companies, railroads, banks, insurance companies, and other urban business interests. Discontent with capitalism itself, however, has been minimal. Rather, farmers have attributed their financial hardships to market imperfections and to the unchecked greed of monopolists and middlemen. Agrarian political activity has centred on increasing farmers' control over the production and marketing of agricultural commodities in an uncertain, shifting world economy. The state is seen as an agent by which such controls can be implemented and the market can be stabilized.

157

After a long period of stability in the 1960s and a boom in the mid-1970s, the 1980s represented years of severe stress in the Mid Western farm economy. Many farmers attributed their economic woes to forces beyond their control. Many argued, for example, that bankers encouraged financial overextension, forcing farmers into recurrent debt crises. The agenda of neo-populist movements in American agriculture can also be regarded as a reaction to the increasing capital-intensity of farming, which has forced many farmers into chronic debt and has prevented young people who lack capital resources from taking up farming. Calls to 'save the family farm' can be viewed as responses to increased mechanization, increased debt stress and capital-intensity, and increased farm size, although the most effective means of countering these perceived or actual threats to family farming is subject to considerable debate (Vogeler, 1981; Tweeten, 1984; Ellig, 1985).

The Nebraska corporate farming amendment

Opposition to corporate intrusion is not new; as early as the 1890s the Populists stressed this issue. For example, a plank in the Farmers' Alliance platform of 1891 read (Hicks, 1931: 433):

> We demand the passage of laws prohibiting alien ownership of land, and that Congress take prompt action to devise some plan to obtain all lands now owned by alien and foreign syndicates, and that all lands held by railroads and other corporations in excess of such as is actually used and needed by them be reclaimed by the government, and held for actual settlers only ...

North Dakota enacted legislation restricting corporate farmland ownership in 1934 (Morrison, 1980). During the 1970s the states of Oklahoma, Kansas, Missouri, Iowa, South Dakota, Minnesota, and Wisconsin followed suit.

Since World War II farm numbers in Nebraska, as elsewhere in the United States, have declined substantially. US Department of Agriculture statistics indicate that the number of farms in Nebraska had declined from nearly 112,000 in 1945 to 60,283 by 1982, while the average farm size rose from 427 to 746 acres. Mid-size, full-time commercial farmers have been hardest hit by the cost–price squeeze: a 1985 survey by the Department of Agriculture indicated that the highest debt–asset ratios and the greatest incidence of negative cash flow were among farms with annual sales between $40,000 and $99,999. However, actual intrusion of corporate ownership of Nebraska farmland has been minimal. In the grain and livestock regions of the American Mid West and Great Plains, farms are highly industrialized and larger than the national average, but non-family corporations are the exception rather than the rule. In Nebraska, for example, it has been estimated that less than 3 per cent of the state's farmland is controlled by non-family corporations (Gregor, 1982; Krause, 1983).

Despite similar conditions, Nebraska's legislature failed to enact anti-corporate farm legislation when neighbouring states did during the 1970s. It was not until the purchase of 33,000 acres of Nebraska land by Prudential Insurance Company sparked fears of further intrusion that activists sponsored a successful petition drive to put an anti-corporate farm initiative on the November 1982 ballot. Initiative 300 is one of the most restrictive anti-corporate laws in the nation, prohibiting non-family corporations and limited partnerships from purchasing farmland or engaging in the production of either crops or livestock.

The political debate over Initiative 300 was intense and bitter. The coalition responsible for the petition drive, the Committee to Save the Family Farm, was composed of liberal, grass-roots farm associations and other sympathetic groups. The Nebraska Farmers' Union was the force behind the coalition and the major source of campaign funds (Lincoln *Journal*, 27 October 1982). Other supporters included the Nebraska Catholic Conference, the Center for Rural Affairs, the American Agriculture Movement, Women Involved in Farm Economics (WIFE), the Nebraska Grange, the Nebraska Pork Producers Association, the National Farmers Organization, and the Nebraskans for Peace (Lincoln *Journal*, 3 November 1982).

The successful placement of Initiative 300 on the ballot gave rise to a well financed counter-coalition that outspent proponents by a margin of ten to one; the $343,000 ultimately generated was a state record for an initiative drive (Lincoln *Journal*, 28 October 1982). The coalition was dominated by the national insurance industry, which provided most of the funds. Prudential Insurance gave an early contribution of $125,000, followed by another $100,000 in a last minute effort to sway the election. The insurance industry was joined by Nebraska banking, real estate, and business interests, industrial farm corporations such as National Farms, and conservative commercial agricultural organizations such as the Farm Bureau and the Nebraska Livestock Feeders Association.

Proponents of Initiative 300 charged that the potential influx of large industrialized and vertically integrated agri-business corporations increased the stress on mid-size, commercial family farms (Lincoln *Journal*, 12 October 1982). They expressed concern that soil conservation would be neglected and that bankrupt corporations would abandon exhausted land. Smaller livestock feeding operations would be threatened by lower prices and higher volumes resulting from tax-loss investment by corporate owners (Center for Rural Affairs, 1985a, b). The timing of the initiative drive in 1981–2 was important to its success. Crop-land prices had boomed during the late 1970s in response to a wave of speculative investment fuelled by the insurance industry. Overseas commodity markets were disappearing to foreign competitors, however, and US crop prices were falling. Local farmers saw their ability to expand curtailed by rising land prices at a time when their capacity to compete with outsiders for a limited land supply was declining. The land price deflation that was to ravage these same areas in the 1980s had not yet set in.

Opponents argued that Initiative 300 would curb capital investment in the state and would force corporate feeding operations to invest in competing states such as Kansas, Oklahoma, and Texas (Lincoln *Journal*, 12 October 1982). Not only would feeding operations leave the state, but desired jobs in agro-industries and food processing (notably meat packing) would be lost as well. At a time when states were clamouring for economic development, Nebraska was seemingly pushing out badly needed capital investment. Interestingly enough, meat packers and processors were conspicuously absent among the opponents of Initiative 300, with the exception of small donations from Iowa Beef Processors.

Initiative 300 passed by a margin of 56 to 44 per cent. However, a county-level map of election results revealed striking contrasts within the state (Fig 10.1). The amendment failed to carry the state's two substantial cities, Omaha and Lincoln. Moreover, it did poorly in the ranching and wheat-producing central and western portions of the state. This opposition was overcome by strong support in the eastern rural areas of Nebraska – counties dominated by mid-size, family-owned grain and livestock farms.

Analysis

Approach

The analysis hypothesized that anti-corporate farm sentiment is related to contradiction experienced as stress during the process of capitalist differentiation. Differentiation by necessity involves the restructuring of power relations among producers. Competition leads to the devolution and dispossession of some producers and the expansion of others. Efforts to preserve or expand agricultural operations can bring about new contradictions and can heighten the process of proletarianization (e.g. the acquisition of more land and livestock might lead to financial overextension and farm foreclosure). However, the threats of dislocation and the options available to those facing economic contradictions are not uniform across time, place, or class locations. At particular times and places some class locations will be subjectively experienced as stressful. When economic contradiction is coupled with unpalatable options to reconcile the stress, farmers often engage in political activity to halt the processes leading to further differentiation. Neo-populist reforms related to the preservation of family farming and the curbing of corporate land ownership attempt to resolve such crises. Historical evidence suggests that neo-populist social struggle mystifies the underlying sources of farm sector stress (Buttel, 1980; de Janvry, 1980). Like many other neo-populist reforms, it is hypothesized that Initiative 300 is an ideological non-counter-hegemonic effort that fails to lead to a substantive restructuring of power relations.

The geography of support for Initiative 300 must be examined with respect to class relations in Nebraska's varied agricultural regions. As Smith (1984) and Harvey (1982) suggest, the process of capitalist development is inherently uneven across space; necessary relations evolve into different outcomes, given varying contingent conditions. The development and penetration of capital fosters particular social relations and spatial structures at given time–space intersections; these outcomes in turn serve to structure the course of further capitalist development. Differentiation has proceeded at varying rates and paths among Nebraska's agricultural regions. It is important to examine support for Initiative 300 in the light of these complexities to delineate those class locations in Nebraska that were most associated with neo-populist reformism plus those locations that offered temporary insulation from economic stress, hence generating lower neo-populist sentiment.

In order to examine these questions, a cross-sectional multiple regression analysis of the voting pattern for Initiative 300 was undertaken. County-level percentages of votes in favour of Initiative 300 served as the dependent variable measuring anti-corporate sentiment. Independent variables were developed from a large initial set indicating capitalist differentiation, production technologies, social relations of production, stress and change in the farm sector, agricultural resources, characteristics of the non-farm sector, and other relevant concepts. Variables expressed in terms of proportions were used whenever possible to capture the polarities that develop during differentiation.

The cross-sectional approach was used to explore the integration of the methods of electoral geography with those of political economy. This analysis is not meant to suggest that empirical regularities define theoretical relationships or imply causality (Sayer, 1984). Rather, causal analysis is required by the finding of empirical regularities which demand, but do not define, explanation. Of central importance is the interpretation of such regularities. The approach assumes that the validity of analysis is based on the critical use of theory and method in the interpretation of results, rather than on the use of specific analytic methods (Chouinard *et al.*, 1984). An interpretation of empirical regularities must take into account the characteristics of places and must understand that empirical regularities are produced by the necessary through the contingent. Accordingly, the analysis concentrates on two agricultural regions of Nebraska, termed the Traditional Corn Belt and Great Plains regions, to provide a concrete (intensive) basis for the interpretation of general (extensive) results (Sayer, 1984: 215).

Methods

The units of analysis were counties, chosen because they are the smallest areal units for which data are readily available. Data for independent indicator variables were derived primarily, although not exclusively, from the US Census of Agriculture (1945 through 1982). Multiple regression methods were applied descriptively; inference statistics were used only as a guide to model development.

Initial exploratory models were estimated by ordinary least squares using individual indicator variables as independent variables. Model estimates were extremely sensitive to the particular sub-set of variables included in a given model, however, indicating a serious technical multicollinearity problem. Multicollinearity can also indicate substantive dependence of individual variables on their contexts. The meaning of particular facts, such as high debt loads, cannot be separated from the context in which they occur. Thus, high debt per farm is associated both with debt stress and with large, capitalist farms. Such problems are consistent with the realist critique of cross-sectional quantitative analysis articulated by Sayer (1984).

Factor analysis was applied to reduce the set of independent variables, replacing the large set of highly correlated independent variables with a smaller set of factors containing most of the information in the larger data set (Table 10.1). The factors were then used as independent variables in the regression analysis. The factor analysis not only simplified interpretation and resolved the technical problems associated with multicollinearity, but it also increased the coherent context dependence of the analysis. Relationships were identified in terms of highly correlated sets of variables, decreasing sensitivity to purely contingent relationships between individual variables.

Table 10.1: Factor definitions in terms of independent variables

Factor name	Independent Variables Factor pattern coefficients (>0.70)[*]	
	Positive coefficients	*Negative coefficients*
Corn Belt/Great Plains	Proportion mid-size farms	Proportion of livestock farms
	Proportion full tenants	Agricultural workers per farm
	Proportion cash grain farms	
Hired services	Value of hired services per farm	On-farm operator
Economic scale	Proportion family corporation farms	Proportion of land operated by individuals
	Average value of land and buildings per farm (sales > $10,000)	Proportion of farms operated by individuals
	Average farm acreage	
	Average net profit	
Intensity	Cattle per farm acre	
	Production expenditure per acre	
	Market value of agricultural products per acre	
	Interest expenditure per acre	
	Hired service per acre	
Long-term consolidation/ disinvestment	Median age of county population	
	Farm consolidation rate 1945–82	
Recent consolidation	Farm consolidation rate, 1978–82	
Part-time	Proportion of part-time operators	Value of family labour per farm
Debt-free	Proportion of farms that pay no interest	
Irrigation	Proportion of harvested acres set aside in government programmes	
	Proportion of farm land irrigated	
	Increase in proportion of farm land irrigated, 1974–8	
Life cycle	Proportion of operators on present farm less than 10 years	
Corporate farms	Proportion of non-family corporate farms	

[*] A Harris-Kaiser oblique principal components analysis. Details available from authors upon request

The revised multiple regression model, estimated using the factors as independent variables, explained 70 per cent of the county-level variance in vote proportion for the amendment (Table 10.2). Thirteen of the 93 Nebraska counties, chiefly urban counties and the core counties of the Sand Hills, were eliminated from the quantitative analyses on the basis of regression influence diagnostics (Wrigley, 1983). Such counties exercise undue influence on the model fitting process; their inclusion would misrepresent relationships among variables in a linear model. The similarity of counties within each of the Great Plains and Traditional Corn Belt regions used in the interpretation was verified using cluster analysis.

Table 10.2: Multiple Regression Analysis for Nebraska Vote on Initiative 300

			Regional Median Value	
Factor	*Beta coefficient*	*P*	*Traditional corn belt*	*Great Plains*
Corn Belt/Great Plain	0.040	0.733	0.60	-1.65
Hired services	-0.062	0.444	-0.75	-0.10
Economic scale	-0.542	0.0001	-1.05	0.75
Intensity	0.277	0.014	0.65	-1.15
Long-term consolidation/ disinvestment	0.497	0.0001	-0.60	-0.60
Recent consolidation	-0.259	0.004	0.05	-0.95
Part-time	-0.163	0.037	-0.95	0.05
Debt-free	-0.086	0.304	-0.25	0.25
Irrigation	0.199	0.037	-0.80	-1.00
Operation age	-0.024	0.740	0.95	-0.05
Corporate farms	0.171	0.034	-0.035	0.45

$N = 80. F = 14.52. R^2 = 0.70$

Discussion

A striking disparity in support for the initiative exists between the two agricultural regions of Nebraska that we call the Traditional Corn Belt and Great Plains regions (Fig. 10.1). Whereas 56 per cent of the voters state-wide supported the initiative, median support was 66 per cent in the Traditional Corn Belt, one of the centres of support for the initiative, but only 51 per cent in the Great Plains. Both regions are rural with little urban population, but are distinct in terms of biophysical environment, agricultural production technologies, the social relations of agricultural production, and the extent and form of differentiation.

Although differences in support for the initiative are associated with variation in production technologies and capitalist development, the strongest associations reflect complexities in the differentiation process. Capitalist transformation is inherently uneven in time and space, proceeding along various routes at differing rates according to the opportunities provided by the material and social histories of places. Support for the initiative is strongest where particular class locations face unpalatable options to acute economic stress; the various routes to proletarianization are not equally feared.

Material basis. The Traditional Corn Belt and Great Plains agricultural regions represent core areas within Nebraska of historical corn belt and great plains production technologies. Although large-scale livestock feeding and irrigated feed-grain operations are found in both regions, most crop land is not irrigated and most agricultural operations represent more traditional forms.

The Traditional Corn Belt is an area of level and gently sloping terrain with deep, well-to-moderately drained fertile prairie mollisols. Precipitation (500–650 mm/year) is adequate for corn and soybeans in most years without irrigation. Irrigation is minimal although increasing (USDA, 1981). Seventy per cent of the land is cropped, mostly to soybeans, corn, and other feed-grains. A large proportion of feed-grains produced are fed to livestock, both cattle and hogs, on the farm where they are produced, and both livestock and feed-grains are major marketed farm products. Agricultural receipts, production expenses, labour, and land values are high per acre. The intermittent labour requirements of crop production are coupled with the regular attention needed in livestock production and feeding. Because labour is applied consistently throughout the production cycle, near equality exists between labour time (e.g. the amount of time labour is applied in production) and production time (e.g. the length of the production cycle).

The Great Plains region consists of the margins of the Nebraska Sand Hills, where excessively drained soils on stabilized sand dunes interpenetrate and grade into the more fertile coarse textured prairie mollisols of the surrounding regions. Slopes are steeper than in the Traditional Corn Belt; precipitation (425–575 mm/year) is lower and more variable, more frequently limiting crop production. Most land is devoted to extensive cattle ranching on native grass rangeland; hay is produced in the stream valleys, and dryland wheat dominates on the better soils of greater extent. Centre-pivot irrigation of corn on sandy soils increased dramatically in the 1970s, but still represents a small proportion of total crop land. Agricultural receipts, production expenses, labour, and land values are significantly lower per acre than in the Traditional Corn Belt. Labour inputs are concentrated around critical operations at distinct points in the wheat and range animal production cycles.

Class structure. The two regions are distinct in terms of the existing degree of differentiation of the class structure. Production in the Traditional Corn Belt is dominated economically by the mid-size family farm closer to the petty commodity producer ideal. Operations are characterized by on-farm, full-time operators and reliance on family labour; hired labour and hired services per farm are low, and part-time operators are less common, due in part to difficulties in reconciling the higher labour demand of intensive livestock production with obligations of off-farm employment. A greater proportion of all farmland is rented or leased, primarily to part-owners. The full- tenant farm is more common, often serving to facilitate intergenerational land transfer in ethnic communities where the preservation of family land is highly valued.

The Great Plains region is more highly differentiated into capitalist and proletarian classes. Large-scale, highly profitable, essentially capitalist farms dominate economically. The large family-owned cattle ranch or dryland wheat operation is frequently operated as a family corporation; the amount of land operated by either individuals or families is significantly smaller than in the Traditional Corn Belt. Hired labour and hired services are greater, even after taking into account the tax advantages accruing to the family corporation that 'hires' family labour. Low sales, part-time farms predominate numerically, however. More than 50 per cent of all farms are small, with less than $40,000 in marketed agricultural receipts. The average age of farm operators is high suggesting that these small, part-time farms are likely to disappear when the current operator dies or retires and the land is purchased by other operators trying to expand their holdings.

Stronger support of Initiative 300 in the Traditional Corn Belt suggests that family farms identified with petty producer ideals are associated with neo-populist ideological responses. These differences in the level and form of capitalist development are captured most fully by two factors: Corn Belt/Great Plains and Hired Services. The Corn Belt/ Great Plains factor distinguishes between the two regions more completely than any other, capturing both the differences in production technologies (e.g. cattle ranching v. feed-grain/livestock production) and in capitalist differentiation of the social relations of production (e.g. mid-size farms dependent upon family labour v. large farms relying upon hired labour). Hired services typically include a high labour component, often reflecting the use of neighbouring farmers' equipment and labour. These two factors are related to the anti-corporate farm vote in expected directions. Where family labour on mid-size farms closer to the petty commodity producer ideal predominates, the vote for the initiative was higher. Where the class structure is more fully differentiated and hired labour and hired services are greater, anti-corporate sentiment was lower.

Contingent routes. The two factors relating most closely to class structure – the Corn Belt/Great Plains and Hired Services factors – are not most strongly associated with anti-corporate sentiment in the regression analysis. The reasons may lie in the complexity and contextuality of the differentiation process. Different routes, or expressions, of capitalist development will be manifested in different places as a result of various biophysical bases, production technologies, and social histories. Other factors than those directly describing class structure may more adequately capture these complexities and contextualities. In other words, at a given time, places are at different junctures along different routes in the capitalist differentiation process. An election is an expression of such a point in time. In the present case, factors differentiating between routes are more closely associated with anti-corporate sentiment than those describing class structure. These include factors describing contingencies such as the scale and intensity of agricultural production, time patterns in consolidation rates, and proletarianization through off-farm employment rather than debt.

The factors describing the Economic Scale and Intensity of agricultural operations discriminate strongly between the Traditional Corn Belt and Great Plains regions and are among the most important correlates of anti-corporate farm sentiment. The juxtaposition of these two factors clarifies relationships between anti-corporate sentiment, production technologies, and stressed class locations. The Traditional Corn Belt is characterized by higher intensity and smaller enterprise scale consistent with its more intensive corn/livestock production technologies and less differentiated class structure. Mid-size commercial scale is combined with high production, labour, and land costs that reduce profit per farm, despite high productivity per acre. This category of farming enterprise was widely identified as particularly subject to severe financial stress during the late 1970s and early 1980s when crop and farmland prices deflated. The close correlation of smaller scale plus higher intensity with vote for Initiative 300 suggests that it is not degree of capitalist differentiation, *per se*, that is associated with neo-populist anti-corporate sentiment, but rather contingent stresses facing particular class positions in particular locales during differentiation.

Farm consolidation rates are related to the rate of capitalist differentiation and can be expected to vary in space and time according to the contingencies associated with place. Two factors, both strongly related to anti corporate sentiment, measure rates of farm consolidation: Long-term Consolidation/Disinvestment and Recent Consolidation. The first factor emphasizes farm consolidation rates between 1945 and 1982 and is correlated with median age of the county population, suggesting high rates of out-migration as young adults are those most likely to migrate. The second emphasizes 1978–82 farm consolidation rates, the interval immediately preceding the anti-corporate initiative drive.

For Nebraska as a whole, long-term farm consolidation rates are high and are one of the most important correlates of anti-corporate farm sentiment. High rates of out-migration, depopulation, and disinvestment in rural communities have been stressful for rural families and rural communities. However, both the Traditional Corn Belt and the Great Plains regions have weathered the changes in farm structure of the past fifty years more successfully than much of the rest of rural Nebraska. Thus, although long-term farm consolidation and community disinvestment can help explain the high overall vote for the initiative in these two regions, they cannot explain differences in the vote between the Traditional Corn Belt and the Great Plains.

Rates of farm consolidation from 1978 to 1982 were higher in the Traditional Corn Belt than in the Great Plains, consistent with reports that the mid-size commercial farm is facing the greatest economic stress. Counter- intuitively, however, this factor is associated with lower anti-corporate farm sentiment. This apparent contradiction is resolved by the opportunities for expansion resulting from an increase of land on the market. In the short term, perceived and actual opportunities for expansion may override the threat of competition. Since 1978 consolidation in the Traditional Corn Belt has been facilitated by the speculative activity of outside investors in prime crop lands. Rising prices brought land on to the market; purchases by outsiders increased land available for lease or rent. The consequent high rates of tenure change facilitated consolidation. The long-term Farm Consolidation/ Disinvestment factor more fully takes into account technological changes and the resulting consolidation imperatives imposed on farmers and farm communities; technological change and the expression of its consequences are long-term processes.

Farmers who lose out in the differentiation process face two options: either (1) leave the farm; or (2) keep the farm, becoming a 'small' farmer as neighbouring farms expand. Those who leave take their votes with them, and the effect of their experience on the community is captured by the Farm Consolidation/Disinvestment factor. The subjective stress of the differentiation process on the farmer may be related to the 'success' of the small farm as a class position that allows a farm family to maintain, temporarily or permanently, a rural farm life style. Two factors in the analysis relate to the likely 'success' of such a transition from a non-viable mid-size to a sub-commercial small farm: the Part-time and Debt-free factors. Higher proportions of part-time and debt-free farms are found in the Great Plains region and are associated with lower anti-corporate sentiment.

The greater prevalence of part-time farms in the Great Plains may arise both from the wider availability of off-farm work in that region and from the greater compatibility of production technologies in the Great Plains with part-time farming. Larger capitalist farms in the Great Plains have a high hired labour and hired services component, offering part-time work to the smaller rancher, even in the absence of non-agricultural jobs. The less intensive labour requirements of cattle ranching and dryland wheat production provide greater opportunity for off-farm work to the small rancher than the more regular labour requirements characteristic of the intensive corn/livestock technologies of the Traditional Corn Belt. Because the production costs of ranching are low, part-time farming is more profitable as well as more convenient in the Great Plains than in the Corn Belt.

The mid-size commercial Traditional Corn Belt farm can rarely operate without debt because production and land expansion expenses are large. The low intensity mid-size ranch in the Great Plains with lower production expenses is better able to finance farm operations without debt. The marginal mid-size operation in the Traditional Corn Belt is, therefore, more likely to be under debt stress or to face dispossession rather than transition to a less intensive, part-time operation. These differences are perpetuated beyond the transition from commercial to sub-commercial; the viability of a debt-free sub-commercial farm is likely to be greater than that of one with debt. In sum, the marginal mid-size operation in the Traditional Corn Belt faces greater contradictory pressures because the small farm option is not as available: a part-time operation is less feasible and the debt constraint is likely to be greater. Maintaining a farming life style while making a gradual transition to wage labour is more difficult, creating 'succeed or fail' consequences for the marginal mid-size Corn Belt operator. Such an 'either/or' dichotomy will be experienced as more stressful than a set of options where intermediate choices are available.

Ideology, social struggle, and social change

The intention of Initiative 300 was to restrain the forces that threaten the viability of family farm production. Proponents drew upon appealing images of farmers and farming that have been pervasive in American history. Although they were faced with a powerful, well financed opposition, farmers under economic stress were able to organize and engender change through the electoral system. In this sense, Initiative 300's passage may reflect a 'successful' social movement. Gramsci (1971) suggests, however, that progressive social change entails more than electoral victories and reforms; it requires the dismantling of ideologies hegemonic in society and the restructuring of relations of power. 'Successful' social movements challenge meanings, values, and power relations of the *status quo* and set forth reforms that directly attack the structural forces that are generating discontent. What becomes central in the case at hand is a comparison of the intent of the reform and its ability to reduce the stresses facing Nebraska family farmers. The question here is whether neo-populist anti-corporate reformism mystifies the actual structural sources of farm sector stress so as to diffuse discontent rather than to focus struggle in a progressive way.

The relationship between the provisions of Initiative 300 and the actual sources of farm sector stress must be examined. In order to be effective, the provisions must be as comprehensive and broadly defined as the sources of threat. There are three sources of corporate threat facing Nebraska producers: local corporate *land ownership*; the scale of corporate operations; and structural changes in the meat-packing industry. By banning non-family corporations and limited partnerships from owning farmland and engaging in agricultural operations in Nebraska, Initiative 300 controls the first threat, but it only partially addresses competition stemming from the scale of corporate operations and fails utterly to deal with drastic changes in the structure of the meat-packing industry.

The Corporate Intrusion factor, measuring the county proportion of non-family corporate farms, is associated with higher support for Initiative 300. This indicates a basis for a relationship between the local visibility of corporate threat and anti-corporate sentiment. The association in the model between local corporate farms and anti-corporate sentiment is not particularly strong, however. Actual corporate land ownership and corporate factory farms are proportionately rare in Nebraska. The incidence of true corporate farming is variable across the state but does not differ greatly between the two regions. Nor is the proportion of corporate farms a good measure of corporate land ownership or competition for land. Although corporate land ownership is low, it may account for a much larger proportion of land transferred, especially in areas of high quality crop lands such as the Traditional Corn Belt. One of the most severe constraints to mid-size commercial family farms is the limited opportunity for expansion posed by low rates of land ownership turnover combined with the high cost of such land. Corporate land ownership may provide significant competition for land available for expansion.

The cost efficiencies, volumes, and locational independence of large-scale corporate farms provide a second threat to the operations of smaller-scale producers. These efficiencies of scale not only pose a competitive threat to producers in the immediate vicinity but affect all producers within that agricultural sector. Curbing non-family corporate land ownership and agricultural operations in the state has little force in dealing with such external competitive pressures. Large corporate operations such as National Farms in Atkinson, Nebraska, with the capacity to feed 300,000 to 350,000 hogs per year, achieve scale economies sufficient to threaten smaller producers not only in the vicinity but throughout the Mid West. The autonomy of individual farming areas and the effectiveness of local policy responses such as Initiative 300 are also greatly constrained by the locational independence and mobility characteristic of large-scale agro-industries. Vertically integrated corporate giants in the food processing industries separate stages of the production process. For example, meat processors might produce feeder cattle in Arkansas, fatten them in Nebraska, and slaughter the animals in Iowa. Economies of scale of corporate farms and agro- industries hinder the ability of local controls such as Initiative 300 to live up to their intent. While Initiative 300 bars the formation of new non-family corporate farms in the state, some corporate farms remain owing to 'grandfather clause' provisions, and competition from corporate farms in surrounding states still exists. This secondary form of corporate threat would be expected to be captured by factors measuring smaller scale, high intensity, family livestock production but not by a factor measuring local corporate expansion.

Nebraska producers face a third corporate threat: increasing concentration and consolidation in the meat-packing industry. Structural ties between the oligopsony power of meat-packers and individual producers, such as forward contracting relationships, are not addressed in the Nebraska amendment. Substantive reforms must examine producers within the broader organization of the food industry. The trend towards fewer, larger producers in livestock feeding is induced by the structural changes in the meat-packing industry, and this will continue unabated by restrictions on corporate livestock feeding or corporate land ownership.

Skaggs (1986) has examined the rising significance of large-scale feed-lot operations in the Plains and Mid West and intensified competition and consolidation in the meat-packing industry, an upsurge 'that portends a return to the days of the beef trust' (p. 215). Three companies, ConAgra, Excel, and IBP, are rapidly gaining market control in beef and pork processing. The repercussions of these changes include the closure of antiquated facilities, drastic reorganization of the work process, the emergence of modern facilities in non-unionized areas, and frequent, sometimes violent, labour struggles (Lauria, 1986). For example, Hormel recently closed its hog slaughter operation in Ottumwa, Iowa, laying off 500 workers making about $11 an hour. Excel plans to reopen the plant paying 800 workers $5.80 per hour, with a significantly higher production capacity.

Economies of scale resulting from this concentration push livestock feeding to fewer, larger producers, lowering meat prices. Larger buyers have strong preferences for larger producers who can ensure lower transaction costs and more uniform quality. As in poultry production, this favours large-scale corporate farms and contract production by smaller, 'family' farmers under conditions where the farmer assumes the risks of meeting the buyers specifications (Wilson, 1986).

Initiative 300 does not restrict contract production and has little impact on the locational independence necessitated by vertical integration in the meat-packing industry. Such restrictions ignore the indirect dispossession associated with contract farming. Contracting is the greatest threat facing mid-size producers, but exploitation is masked by the appearance of independence. Although operators still own the means of production (confinement sheds, equipment, land, etc.), decisions about what, how much, and when to produce are made elsewhere (Davis, 1980).

The growth of contracting in the production of cattle and hogs suggests an alternative interpretation of the Economic Scale and Intensity factors in terms of the distinction between labour time and production time. In some productive sectors there is a lack of correspondence between labour time – periods when labour is actually applied – and production time – periods when unfinished commodities are 'abandoned to the sway of natural processes' (Mann and Dickinson, 1978: 472). Examples of the latter include the natural, chemical, and physiological processes required in the maturing of crops and the reproduction of livestock. These intervals do not create value or surplus value, but are necessary to the production of finished commodities.

Because of differences in labour time and production time, the perishability of crops, and the uncertainties of the weather, agricultural production is seemingly incompatible with the requirements of capitalist production. Yet some agricultural sectors exhibit a greater variance in labour requirements during the course of the production cycle than others (e.g. dryland wheat farming has a higher variance than intensive livestock/crop production). Intensive livestock/crop production, characterized by regular labour inputs during the production cycle, has a more nearly equal ratio of labour time to production time that makes this sector more attractive to capitalist intrusion. In the Traditional Corn Belt, Economic Scale is low and Intensity is high. The reverse is true of the Great Plains, implying that high support is associated with economically vulnerable farms with higher ratios of labour time to production time. It is likely that there will be more capitalist development in the intensive crop/livestock production sector, and such intrusion will continue despite measures such as Initiative 300. However, investment will be deflected away from corporate land ownership to more indirect forms of control like contracting.

An examination of the coalitions opposing the initiative is

illustrative. The 'No On Initiative 300' coalition included insurance and financial interests and corporate farms such as National Farms, all of which had a direct stake in corporate land ownership. Only one meat packer contributed to quell the initiative, and its contributions were greatly overshadowed by the $280,000 given by major insurance companies (Lincoln *Journal*, 28 October 1982). Such involvement was unnecessary because Initiative 300 did not fundamentally threaten the meat packers' investment decisions or relations with livestock producers and feeders.

Initiative 300 channelled discontent while leaving unleashed processes leading to increased scale and concentration in the system of food and fibre production. The choices open to Nebraska farmers were not significantly altered by the form of protest embodied in the initiative. Ideologically, Initiative 300 served to maintain the symbol of petty commodity production and in the process limited the scope and revolutionary potential of the debate.

Summary

Much as analysts of the class relations of agriculture suggest, differentiation is not unilinear. It is a complex, uneven process riddled with contingency, and farm sector stress can be resolved in different ways. Comparatively, the two regions reflect distinct paths of differentiation and separate resolutions of economic contradiction.

Initiative 300 is a contemporary version of a long line of populist reforms in the Mid West and Great Plains. Strongly laden with ideological notions about the family farm and the product of good intentions, Initiative 300 ignores several structural forces that affect Nebraska producers by focusing solely on the issue of corporate farming and corporate land ownership. In this respect, neo-populist reformism mystifies the actual forces of farm sector stress. Support for and opposition to the initiative can be understood with regard to tensions and tolerable options facing particular, often contradictory, class locations.

References

Bellah, R. et al. (1985) *Habits of the Heart: Individualism and Commitment in American Life*. Berkeley, Cal.: University of California Press.

Buttel, F. H. (1980) 'Agriculture, environment, and social change: some emergent issues', in F. H. Buttel and H. Newby (eds.) *The Rural Sociology of the Advanced Societies*. Montclair, N.J.: Allenheld Osmun, 453–88.

—— and Flinn, W. L. (1976) 'Socio-political consequences of agrarianism', *Rural Sociology* 41, 4: 473–83.

Center for Rural Affairs (1985a) 'Corporate hog contracts spread', *CRA Newsletter*, November: 2.

—— (1985b) 'False charges against 300 rebutted', *CRA Newsletter*, January: 3–4.

Chouinard, V., Fincher, R. and Webber, M. (1984) 'Empirical research in scientific human geography', *Progress in Human Geography 8*, 3: 347–80.

Craig, R. A. and Phillips, K. J. (1983) 'Agrarian ideology in Australia and in the U.S.'. *Rural Sociology* 48, 3: 408–20.

Davis, J. E. (1980) 'Capitalist agricultural development and the exploitation of the propertied laborer; in F. H. Buttel and H. Newby (eds.) *The Rural Sociology of the Advanced Societies*. Montclair, N.J.: Allenheld Osmun, 133–54.

de Janvry, A. (1980) 'Social differentiation in agriculture and the ideology of neo-populism', in F. H. Buttel and H. Newby (eds.) *The Rural Sociology of the Advanced Societies*. Montclair, N.J.: Allenheld Osmun, 155–68.

Ellig, N. (1985) 'Issues in the analysis of contemporary farm protest', *Agriculture and Human Values* 2: 44–8.

Goldman, R. and Dickens, D. R. (1983) 'The selling of rural America', *Rural Sociology* 48, 3: 585–606.

Goodwyn, L. (1986) 'Populism and powerlessness', in H. C. Boyte and F. Riessman (eds.) *The New Populism: the Politics of Empowerment*. Philadelphia, Pa.: Temple University Press, 19–29.

Goss, K. F. Roidefeld, R. D., and Buttel, F. H. (1980) 'The political economy of class structure in U.S. agriculture: a theoretical outline', in F. H. Buttel and H. Newby (eds.) *The Rural Sociology of the Advanced Societies*. Montclair, N.J.: Allenheld Osmun, 83–132.

Gramsci, A. (1971) *Selections from the Prison Notebooks of Antonio Gramsci*. (ed. and trans.) Q. Hoare and G. Howell, New York: International Publishers.

Gregor, H. F. (1982) *Industrialization of U.S. Agriculture*. Boulder, Colo.: Westview Press.

Harvey, D. (1982) *The Limits to Capital*. Oxford: Oxford University Press.

Hicks, J. D. (1931) *The Populist Revolt*. Minneapolis, Minn.: University of Minnesota Press.

Kautsky, K. (1980) 'Summary of selected parts of the agrarian question', in F. H. Buttel and H. Newby (eds.) *The Rural Sociology of the Advanced Societies*. Montclair, N.J.: Allenheld Osmun. 39–82.

Krause, K. R. (1983) *Corporate Farming: Importance, Incentives and State Restrictions*. Economic Research Service, Agricultural Economic Report No. 506, Washington. D.C.: USDA.

Lauria, M. (1986) 'Toward a specification of the local state: state intervention strategies in response to a manufacturing plant closure', *Antipode* 18, 1: 39–65.

Lenin, V. I. (1899)) *The Development of Capitalism in Russia*. repr. 1977, Moscow: Progress Publishers.

Lincoln *Journal*, (1982a) 'Anti-initiative 300 corporate gifts blasted', 28 October: 1.

—— (1982b) 'Donation record set in initiative 300', 27 October: 41.

—— (1982c) 'Initiative 300 debaters agree little on implications', 12 October: 8.

—— (1982d) 'Initiative 300 passed by voters', 3 November: 1.

174

MacPherson, C. A. (1962) *The Political Theory of Possessive Individualism: Hobbes to Locke.* Oxford: Clarendon Press.

Mann, S. A. and Dickinson, J. A. (1978) 'Obstacles to the development of a capitalist agriculture', *The Journal of Peasant Studies* 5, 4: 446–81.

Marx, K. (1852a) *The Eighteenth Brumaire of Louis Bonaparte*, repr. 1967, New York: International Publishers.

—— (1852b) *Capital.* repr. 1979, New York: International Publishers.

Mooney, P. H. (1978) 'The transformation of class relations in Wisconsin agriculture 1945–1982', Ph.D. Dissertation, University of Wisconsin-Madison.

—— (1982) 'Labor time, production time, and capitalist development in agriculture: a reconsideration of the Mann-Dickinson thesis', *Socialogia Ruralis* 22, 314: 279–91.

—— (1983) 'Toward a class analysis of Midwestern agriculture', *Rural Sociology* 48, 4: 563–84.

—— (1986) 'The political economy of credit in American agriculture', *Rural Sociology* 51, 4: 449–70.

Morrison, F. (1980) 'State corporate farm legislation', *University of Toledo Law Review* 7: 961–97.

Sayer, A. (1984) *Method in Social Science: a Realist Approach.* London: Hutchinson.

Skaggs, J. M. (1986) *Prime Cut: Livestock Raising and Meatpacking in the United States 1607–1983.* College Station. Tex.: Texas A and M University.

Smith, N. (1984) *Uneven Development.* Oxford, Basil Blackwell.

Tweeten, L. (1984) *Causes and Consequences of Structural Change in the Farming Industry.* Washington, D.C.: National Planning Association.

United States Department of Agriculture (1981) *Land Resource Regions and Major Land Resource Areas of the United States.* Washington, D.C.: Soil Conservation Service.

Vogeler, I. (1981) *The Myth of the Family Farm: Agrobusiness Dominance of U.S. Agriculture.* Boulder, Colo.: Westview Press.

Wilson, J. (1986) 'The political economy of contract farming', *Review of Radical Political Economics* 18, 4: 47–70.

Wood, N. (1984) *John Locke and Agrarian Capitalism.* Berkeley, Cal.: University of California Press.

Wrigley, N. (1933) 'Quantitative methods: on data and diagnostics', *Progress in Human Geography*, 7, 4: 567–77.

Chapter eleven

Social transformation and changing urban electoral behaviour

David C. Hodge and
Lynn A. Staeheli

The economic restructuring that characterizes modern capitalism has had profound effects on the social and political structure of society, the spatial structure of cities, and the relationship between them. Attempts to account for these changes have stimulated, if not demanded, the use of theoretic frameworks that more explicitly attempt to interpret urban phenomena through an understanding of the deeper structures of society in which diverse spheres of activity are woven together (Johnston, 1984). Thus this chapter seeks to broaden our understanding of one important urban phenomenon, electoral behaviour, by grounding it firmly in the context of materialist theories of social change. Specifically, following Katznelson (1981), we shall focus on the divergence between production relations (and the politics of the workplace) and consumption relations (and the politics of the home) that characterizes if not distinguishes the American social and political landscape. In particular we will attempt to show how political interests rooted in the politics of consumption shape the nature and outcome of electoral issues differently from political interests rooted in the historic coalition of partisan and production politics.

This thesis is examined by conducting an analysis of electoral behaviour in the Seattle metropolitan area focused both empirically and theoretically on the relationships *among* electoral issues. We first identify the underlying dimensions of political salience that are expressed by the empirical similarities among individual ballots and then relate these dimensions to theories of social change. It is argued that the hegemony of traditional party politics (which have been tied to definitions of production-based class) is being replaced by a more complex set of electoral dimensions that reflect *both* consumption and production interests.

The structure of urban electoral issues

Although individual ballot issues are routinely analysed, there have been remarkably few attempts to understand the relationships among a full range of ballot items at American elections (Guest *et al.*, 1988; Hahn and Kamieniecki, 1987). This is unfortunate in that it is critical to understand not only the motives of voters but also the nature of the issues with which they are confronted. It is important to note, however, that our analysis assumes that ballot items do reflect fundamental social issues, a position subject to considerable debate. Indeed it has been persuasively argued that most of the real issues affecting the social structure of society are determined outside the overt political arena of elections (see, for example, Bachrach and Baratz, 1970), a view which obviously trivializes any attempt to relate electoral behaviour to deeper social structure. While we concede the efficacy of this position with respect to some aspects of political culture, we do not take such a view to be reflective of the primary functions of electoral processes. That is, we assert that elections do have symbolic and real meaning with respect to social struggles in American society. Elections are not simply a constrained activity which by appearance alone legitimates the state (Lukes, 1975) but reflect individual and cumulative decisions that affect fundamental issues of society (Castells, 1978) as well as offering considerable autonomy, rarely exercised, to challenge the nature of the state (Staeheli and Hodge, 1988). Thus it is logical to expect that the dimensional structure of elections in some ways reflects major social cleavages.

Elections consist of two types – candidates and issues. At the national level in the United States elections are formally restricted to candidate elections either through direct election of the President and Vice-president (the electoral college not withstanding) or indirect representation in Congress. It is at the state and local level, at least for the 22 states which provide for their use, where issue elections (referenda, initiatives, bonds, etc.) are most prevalent, although candidate elections and partisan politics obviously play a role as well. Two questions are especially important here. First, there is a need to understand the relationship among the various issue elections, to identify the extent to which they reflect a common structure. Second, there is a need to determine the extent to which issue elections follow the pattern set by candidate, and more specifically party, elections.

Theoretical considerations

There are two characteristics of partisan and issue elections that are critical to our understanding of the relationship between them. First, there is the question of whether or not there is a pattern within each set of elections, that is, whether or not there is a commonality among partisan elections and/or a commonality among issue elections. Since it is well established that overall there is a common pattern to partisan elections, the first question reduces to whether or not there is any commonality among the issue elections. Second, there is the question of how the two sets of issues are related to each other, that is, whether issue elections follow the pattern set by partisan elections or whether they diverge in a systematic fashion. Thus there are three logical possibilities for the relationship between partisan and issue elections: (1) both sets of elections have internal commonality and they are related to each other; (2) issue elections are neither organized with respect to each other nor correlated as a group with partisan elections; and (3) both sets of elections have internal commonality and issue elections systematically diverge from partisan elections. The theoretical explanations tied to each of these possibilities will be discussed at some length since these explanations are critical to an understanding of our final conclusions. For reasons that will be clarified below we have chosen to label these three explanations as: (1) the nationalization thesis; (2) the exceptionalist thesis; and (3) the restructuring thesis.

Nationalization thesis

The first explanation argues that the dimensional structure of election issues is primarily shaped by the structure of national partisan politics. The two-party system is deeply ingrained in American society. Certainly, as argued earlier, it represents the overwhelming way in which political matters are structured at the national level and the state level and it penetrates deeply into most politics as well (with some very notable exceptions where reformist and populist ideals have offered non-partisan candidate elections in form if not in substance). Even though there has been a growth in 'uncommitted' or 'independent' voters, the concept of party politics and the political platforms they represent are deeply etched into the psyche of the average voter. The traditional cleavages between Democrat and Republican include divergent views on the role of government in both economic and personal affairs, the nature of the social contract whereby economic status and the redistribution of wealth and power are identified, and the basic rules by which society is to be governed. The fundamental nature of these traditional issues is class-oriented and, by implication, primarily reflects production relations.

Although class consciousness is not nearly as well developed in the United States as in Europe, the Great Depression was serious enough to produce 'enduring class–party alignments' with the Republican Party characterized as the party of big business, small town residents, and the rich, whereas the Democrats became known as the party of the working person (Hahn and Kamieniecki, 1987: 34). While there is very little support for a simple twofold division of social class in American society, strong evidence exists that most Americans do have some class consciousness that is most clearly tied to economic status and, more specifically, to occupational status (Jackman and Jackman, 1983). Compared to single issue elections, candidate elections embody an often complex set of issues which the voter resolves most easily through the candidate's party identification which, we argue, is principally tied to class as defined in the workplace.

There are two aspects of the evolution of party politics that relate to our expectations regarding the relationship between partisanship and the dimensional structure of elections. First, there is the issue of the extent to which party politics have gone beyond the politics of production relations to embrace the politics of consumption relations. If we accept the rational voter model, it logically follows that political parties adjust to changes in political attitudes and incorporate contentious social issues into their agenda. Indeed, such behaviour has been observed by Johnston (1986) who argues that political parties in Great Britain have realized the need to penetrate into all aspects of political issues (this in spite of their closer connection to class defined in the workplace). If this model holds in the US context, then we expect to see the structural dimensions of electoral issues to be strongly related to partisan contests. Second, it has been argued that a nationalization of voter attitudes has been occurring in which regional variations in voting behaviour have given way to a more uniform national consciousness that is based on social cleavages (a position described but rejected by Agnew, 1987). Within this framework regional variation (and by implication intra-urban variation) is a result of variation in the socio-economic status of voters and not of real behavioural differences between regions. Again, we expect the major thrust of issues to be consistent with that of partisan politics; while patterns of support and opposition to a specific issue may or may not coincide exactly with the general pattern for partisan politics, according to the nationalization explanation the major trend of such issues should follow the cleavages structured by and embodied in partisan politics.

Exceptionalist thesis

A second view of issue-oriented ballots may be termed exceptionalist in that while some of the issues would logically follow traditional lines of political behaviour, other issues might be either unrelated to the national agenda or possibly at odds with the traditional structures of support. Two different perspectives might lead to this result. First, if it were true that most real political issues were filtered out of the electoral process rendering it a trivial action (a view we rejected above), then there would be no necessary reason to expect a clear dimensional structure to ballots.

Rather an unpatterned relationship between and among issue and partisan politics would be expected. Second, it may be argued that issue politics are by their nature primarily isolated from each other. A particular referendum, it might be argued, would either have been subsumed into normal partisan legislative politics or would stand alone as 'single issue' politics (Hahn and Kamieniecki, 1987). A particularly good example of the latter is provided by Kirby (1987) who notes that local administrations (or in many cases local electors) who decide to declare themselves as nuclear free zones do so with no real belief that their actions are effective (nor with any direct connections to classic party lines) but rather do so as a means of expressing opposition to policies of the state which they feel is denied through traditional partisan channels. Another related set of 'exceptional' behaviours may simply reflect a regional idiosyncracy, an issue that is superficial with respect to those issues that are contentious among social classes defined nationally. Such issues would logically be unrelated to the structure of major partisan elections or their roots in social cleavages, and would further be unrelated to each other in any systematic fashion.

Restructuring thesis

A third viewpoint differs sharply from the exceptionalist perspective in that it expects issue-oriented politics to differ from partisan party politics in a logical and consistent pattern. That is, the votes on a large number of issues are generally related to each other and not closely related to partisan voting. It is not difficult to argue that many local elections and issues arise contrary to national patterns expressly because the constituencies supporting those issues are not aligned with the national pattern (the nuclear freeze debate, with its broad bipartisan support and opposition, is a good example). Still we are left with the problem of whether or not there is any commonality among such issues or whether they are independent of each other as well as the national profile. It is one thing to assert that such issues exist independent of partisan political norms, it is another to assert that rebellion is organized along a recognizable conceptual continuum. However, we argue precisely that point by appealing to broader social theory which contends that there are an emerging set of issues which do have a common theme and which do differ from the themes of traditional party politics in a coherent fashion.

We argue that the separation from traditional party politics that may be revealed in issue elections is consistent with the economic and social restructuring that has been occurring in the United States. As this society has evolved into modern capitalism new social cleavages have developed. Significantly, these cleavages are not only tied to classes based on production relations but are also tied to social distinctions in the realm of consumption, with the growth of Fordism following World War II as the focal point of this distinction. The economic enfranchisement of the middle class that was necessary to the creation of consumer markets created conditions of contrary interests. Thus significant parts of the middle class found that their allegiance was being torn between the politics of the workplace (as they affected one aspect of their class identity) and the politics of the home (see Castells, 1978, and Katznelson, 1981) where, among other things, mortgages were redefining their social position and political behaviour (Dunleavy, 1979; Saunders, 1984; Pratt, 1986). One of the most dramatic statements of this separation of consciousness is provided by Bunge (1977: 64):

'Part of life is spent at home, in urban neighborhoods ... Capital encourages profligacy and indebtedness at home and in contrast dedication and restraint at the workplace ... Working class consciousness is diluted by this separation. Working class interests are cemented in the crucible of daily production ... and are unravelled just as fast ... at the point of reproduction'.

(As cited in Burnett, 1984: 39).

The rise of the 'welfare state' that accompanied Fordism is an additional part of the consumption landscape, in this case collective consumption (Castells, 1978; Pinch, 1985). Political issues related to collective consumption, which have been favourite targets for elections, ironically have tended to emphasize the protection of private property and its accompanying spatial forms, i.e. suburbanization (Gottdiener, 1985). Thus it follows that the common structure of issue elections ought to deviate in significant and systematic fashion from partisan elections since the former stress consumption issues while the latter reflect more on issues of production relations.

Patterns in electoral issues

In order to test these three alternative explanations data were gathered for 26 partisan candidate and 36 issue ballots cast between 1978 and 1984 in King County, Washington. In many ways King County represents an ideal testing ground for this analysis. Not only is Washington one of only 22 states that permit the initiative process, but use of the initiative process has been substantial there. The data were provided by the King County Elections Department and were reported by precincts. For each election precinct totals were aggregated upwards to census tracts, with split precincts assigned to the census tract which contained the largest population portion of the precinct. There were on average about eight precincts for each tract. For candidate elections variables were defined as the percentage of voters who voted for the Democratic candidate. For issue elections variables were defined as the percentage of voters who voted 'yes' for the issue. The list of candidate elections and issues (including a descriptive label) is presented in Table 11.1. (A more detailed description of the issue ballots can be found in Guest *et al.*, 1988.)

The relationship between these separate ballot issues was analysed via a factor analysis with oblique rotation in order to generate the best picture of the electoral structure without distorting the identity of individual dimensions by forcing statistical independence. If the nationalization explanation is correct, then we would expect to see substantial similarity in the reaction by voters to both partisan and issue elections (as measured by comparable patterns of factor loadings on a major factor). If the exceptionalist explanation is correct, then the pattern should be more confused with very little systematic covariation within the issues or between the issue ballots and the partisan elections. If the restructuring explanation holds, then we would expect that at least one dimension should emerge that is significantly different from the general pattern of support for partisan candidates.

The results of the analysis are remarkable in terms of the clarity of the dimensions that emerge (Table 11.1). Two factors account for more than 75 per cent of the variance with the residual fragmented among a number of nondescript dimensions. Not only are these two factors dominant in the factor solution but they are also distinct from each other (only a correlation of 0.025 between them). Even more noteworthy is the composition of the factors, which lends considerable support to the restructuring explanation. Almost all partisan and a few issue elections load heavily on one dimension while most issue and few partisan elections load heavily on the other dimension.

Factor I is clearly a measure of partisan support. Most of the candidate loadings are in excess of 0.90 and only for two elections, the US Senate race in 1982 and the Congressional races in 1980, are the loadings on the second factor higher than the loadings on the first factor. In the 1982 Senate race, Henry Jackson, a long-time Senate Democrat who was popular among Republicans because of his conservative views on defence, obscured party distinctions in his usual one-sided victory. The mixed pattern for the three Congressional seats in 1980 is in part a function of the fact that the variable includes more than one candidate but also it is likely a reflection of the overwhelming Reagan victory, which influenced many of the other partisan races that year. Other than these two elections, then, the first factor portrays well the most fundamental cleavage in American politics, that of partisan politics whereby patterns of support for candidates are strongly related to their party affiliations.

The second dimension is distinguished by its affiliation with non-partisan issue politics. Although the correspondence between these two sets is not strictly one-to-one, in general non-partisan issues load heavily on factor II, which comprises non-partisan measures almost exclusively. This gives strong credence to the restructuring explanation in that factor II represents a dimension of electoral behaviour that focuses on issues that are (1) united in a common outlook and (2) independent of traditional party cleavages. We have chosen to label this factor American ideology because it embodies issues that strike uniquely at the heart of American society. The extent of and reasons for American 'exceptionalism' have been hotly debated, but we follow Katznelson (1981) who argues that politics in the United States are different from those in Europe and that it is the separation of issues in the workplace from issues of the home that is especially important. It is particularly important to note that, with respect to the home, issues related to consumption figure prominently, a reflection, we argue, of the importance of consumption to the peculiar historical context of the United States as well as of changes that are part of the transformation of modern capitalism.

As James Bryce noted in his nineteenth century masterpiece *The American Commonwealth* (1894), Americans are a people of paradox who are 'changeful' but at the same time 'conservative'. This inherent contradiction in the American ethos between principles which so strongly support the rights of individuals, etc. and the conservatism that is reflected in the limits set to those rights is, in the twentieth century, in no small way tied to protecting the home environment and all the consumption patterns that are implied therein. Americans are more liberal at the highest levels of abstraction and more conservative when faced with the realities of the everyday environment. For example, Americans champion the rights of minorities at the national level but institute every possible spatial rule to ensure that associated with the property rights of suburban location are economic and ethnic homogeneity. Americans champion the entrepreneur but erupt bitterly against development proposals that alter the physical or social environment around their homes. The political divisions surrounding the consumption issue are in many ways highlighted in progressive politics, which, while not immune from issues of personal consumption, have stressed an interest in the common good, as opposed to either party politics, which stressed the interests of specific classes or more importantly boss rule, which was the ultimate machine for controlling consumption. Thus there exists a long-standing tension between the emphasis on private consumption (or public consumption for private gain) and political conservatism and the emphasis on progressive politics rooted in the liberal values of society that, at least in appearance, champion public consumption and the common good and have done so outside partisan platforms.

Environmental issues	Factor I	Factor II
I383. Ban Import of Radioactive Waste, 1980	0.67	0.02
KCPropl. Expand Park and Recreation Facilities, 1982	0.07	0.93
1414. Deposit on Beverage Containers, 1982	0.23	0.88
SJR112. Allow Utilities to Finance Energy Conservation, 1983	0.03	0.94
Public services		
KCProp2. Tax to support Emergency Aid Car, 1979	-0.61	0.08
Metro-Sales Tax Increase for Metro Transit, 1980	-0.27	0.83
R38. Bonds to Improve Water Supply Facilities, 1980	-0.12	0.90
R39. Bonds to Improve Waste Disposal Facilities, 1980	-0.11	0.93
KCPropl. Telephone Tax to Improve 911 Service, 1981	-0.62	0.17
KCProp2. Bonds to Improve Jail Facilities, 1981	-0.11	0.92
Government structure		
SJR110. Allow Legislature to Meet Annually, 1979	0.21	0.89
SJR112. Permit Legislators to Assume Other Civic Offices, 1979	-0.28	0.80
KCProp1. County Takeover of Metro, 1979	0.41	-0.14
HJR37. Create Judicial Qualifications Commission, 1980	0.03	0.72
SJR132. Make State Claim to Federal Lands, 1980	0.01	-0.20
KCCA. Permit Government Employees to Hold Public Office, 1981	-0.08	0.70
SJR107. Permit Legislators to Increase Judicial Help, 1981	-0.09	0.82
SJR133. Reduce Number of Signatures for Initiative, 1981	-0.78	0.25
SJR103. Create Redistricting Commission, 1983	-0.13	0.80
Taxes and business		
I62. Limit Government Spending, 1979	-0.69	-0.11
I394. Require Vote on Major Energy Projects, 1981	0.80	0.27
I402. Abolish Inheritance Tax, 1981	-0.50	-0.78
HJR7. Permit Industrial Development Bonds, 1981	-0.43	0.38
I412. Impose Maximum Credit Card Interest Rate, 1982	0.60	-0.08
I435. Increase Corporation Taxes, Reduce Other Taxes, 1982	0.88	0.22
SJRI43. Allow Tax on Benefit from Public Improvements 1982	0.05	0.95
SJRI05. Increase Length of Harbor Leases, 1983	-0.40	0.36
I464. Exclude Auto Trade-ins from Sales Tax, 1984	-0.58	-0.77

A closer examination of the composition of the second factor clarifies the meaning of the American ideology dimension and also explains logical inconsistencies in the correspondence between issues and factor II; some specific issues are aligned more closely with partisan politics than with other non-partisan issues. The issues have been grouped into five major clusters which reflect their general substance: taxes and business, government structure, public services, environment, and social issues. Not counting the cluster of tax and business issues the overall cohesiveness of the issues is impressive with 21 of 27 loading most heavily on factor II. Their common characteristics and relevance to American ideology can be observed by considering each of the major clusters in turn. In addition, inconsistencies (issues loading more heavily on factor I) will be discussed and explained.

We begin with the biggest set of exceptions, the issues labelled *taxes and business*, which we expect to differ from the other four clusters given their importance to traditional partisan politics. In fact six of the nine elections, which include measures limiting government expenditure and increasing corporate taxes, do follow traditional partisan cleavages and load most heavily on factor I. Two issues – the measure to abolish the inheritance tax and the measure to exclude auto trade-ins from the sales tax – loaded heavily on both dimensions, reflecting complex patterns of support and opposition. The measure to allow taxes on the benefits to public improvements was the most noteworthy exception in this set of issues, since it loaded very heavily on factor II with virtually no relationship to factor I. We speculate that voters looked at that measure through classic progressive norms in which the salient issue was the debate over the extent to which private parties should benefit from public largesse.

A second group of issues, *government structure*, is even more easily characterized as relating to progressive causes since 'good government' is a critical part of such movements (Magleby, 1984). Seven of the nine issues, which generally allow for greater participation in political processes, loaded most heavily on factor II and most had high loading values. Two issues from this group were exceptions and loaded most heavily on the partisan dimension. The first exception, a county-wide proposition to have Metro (the authority controlling sewage and transit) taken over by the county, correlated moderately with Democratic responses probably because of the pro-government/anti-government split between Democrats and Republicans. More intriguing is the association between the vote for a measure to reduce the number of signatures necessary for initiatives and Republican support. We speculate that the support for the right to challenge government (which is consistent with Republican views in favour of less government), combined with the historical domination of the state legislature by the Democratic Party may be part of the reason for this outcome. It is also true, however, that many particularly contentious social issues, such as eliminating bussing and gay rights, are usually aligned with conservatives who have had more political success using initiatives, thus leading to their support for the initiative process.

The group of *public service* ballots include a number of issues that are clearly related to traditional forms of collective consumption like sewers, jails, and transport. Four of the six ballots have very high loadings on factor II. The two measures that run counter to this pattern represent a set of unusual circumstances. The measures to provide taxes to improve emergency aid car and 911 response probably followed more traditional partisan patterns of support, since these services were already being provided in the predominantly Democratic city of Seattle. Thus the advantage of these strategic services (for which there is not an equivalent private market alternative) would flow to the predominantly Republican suburbs. It is also interesting to note that, while all the other issues have low loadings on factor I, the direction of the loadings indicates some correlation with Republican voting which would seem to run counter to the usual Republican opposition to increased taxes. However, modest support for the bond issues can be explained by their relevance to stimulating economic growth or improving law enforcement which are typical Republic issues.

With the exception of the Bill to ban the importation of nuclear wastes into the state, all the *environmental issues* were strongly aligned with the American Ideology factor. Whether it was encouraging conservation or providing for farmland preservation and more parkland, the issues of reverence for land and support of the collective good seem to be the common elements linking these votes to that ideology factor. It is further interesting to note that the correspondence between the environmental issues and partisanship is directly opposite that of public services. Overall there is a weak association among these issues, which tend to involve expenditure not directly tied to economic development and/or involving more government control, with the Democratic vote.

More than any other set of issues, those ballots which we have grouped together as *social issues* reflect a base in both of our electoral factors. This makes sense in that these issues involve the limits of personal rights, which is a contentious and often inflamed division between Democrats and Republicans, as well as the rights of consumption and the support for a pluralistic culture (Guest *et al.*, 1988) that are embodied in the American Ideology. It is especially interesting, then, that these issues, which include a measure to restrict abortions, are more closely associated with the American Ideology dimension. Apparently the connection of these issues with the progressive spirit and pluralistic culture embodied in the American Ideology dimension is a better distinguishing agent than is partisanship at least for three of the measures. In at least one case, however, it appears that the link with consumption may be the critical connection. Initiative 350 would have banned the bussing of students to all but the closest or next closest school, which we argue was an attempt to protect the property rights associated with residence.

Conclusion

The primary purpose of this chapter has been to respond to the growing call for the need to ground electoral geography in deeper and more appropriate theoretical contexts. Thus we have attempted to understand the relationships between ballot items by identifying the structure of electoral behaviour and placing it in a proper historical and theoretical context. In this sense we have found an impressive clarity and correspondence with theory in the structure of electoral behaviour in King County. Two dimensions emerged from our analysis of nearly ten years of ballots. One dimension is composed of partisan elections whose dominant theme concerns traditional production relations while the other dimension is oriented towards issue elections whose dominant theme, we argue, is a concern for consumption, both private and collective and oriented strongly towards the home. These results are seen as evidence for Katznelson's (1981) argument that the separation of political interests at work and political interests at home is one of the most distinctive features of American society. Although the roots for this distinction run deep in the origins of American culture we argue that the Fordist economic regime following World War II solidified this division. The economic enfranchisement of labour combined with rapid suburbanization in a culture of private property rights virtually assured the ascendancy of consumption interests that are separate from the issues founded in relations at work.

However, we are left with a number of questions regarding the generalizability of our framework and findings. In the case of King County, we have found a dramatic fit between theory and observation both in terms of the way issues are perceived with respect to each other and in terms of the patterns of social cleavage that give rise to that electoral behaviour. We do not know, for example, the extent to which the institutional arrangements governing elections in Washington affect the overt political agenda. There is also a question regarding the lack of strong partisan political traditions in Washington. Finally, there is also a question of the extent to which Seattle is typical of US metropolitan areas. While in the largest sense Seattle appears to encompass the usual social and geographic distinctions common to US metropolitan areas, it is a younger city without some of the contrast present in older more sharply differentiated cities.

Still, we are encouraged to believe that in as much as Seattle is embedded in the same regime of modern capitalism, processes observed there are similar to those found elsewhere in the United States. While institutional and other differences may result in a somewhat different empirical outcome, the logic of the underlying social theory we believe provides a compelling framework for shaping the nature of the interpretation of urban electoral geography.

References

Agnew, J. A. (1987) *Place and Politics: the Geographical Mediation of State and Society*. Boston, Mass.: Allen and Unwin.

Bachrach, P. and Baratz, M. (1970) *Power and Poverty: Theory and Practice*. New York: Oxford University Press.

Bryce, J. (1894) *The American Commonwealth*. New York: Macmillan.

Bunge, W. (1977) 'The politics of reproduction: a second front'. *Antipode* 9, 2: 6–76.

Burnett, A. (1984) 'The application of alternative theories in political geography', in P. Taylor and J. House (eds.) *Political Geography: Recent Advances and Future Direction*. London: Croom Helm.

Castells, M. (1978) *City, Class, and Power*. London: Macmillan.

Dunleavy, P. (1979) 'The urban basis of political alignment', *British Journal of Political Science* 9: 409–43.

Gottdiener, M. (1985) *The Social Production of Urban Space*. Austin, Tex.: University of Texas Press.

Guest, A., Hodge, D., and Staeheli, L. (1988) 'Industrial affiliation and community culture: voting in Seattle', *Political Geography Quarterly* 7: 49–73.

Hahn, H. and Kamieniecki, S. (1987) *Referendum Voting: Social Status and Policy Preferences*. New York: Greenwood Press.

Jackman, M. R. and Jackman, R. W. (1983) *Class Awareness in the United States*. Berkeley, Cal.: University of California Press.

Johnston, R. J. (1984) 'The political geography of electoral geography', in P. J. Taylor and J. House (eds.) *Political Geography: Recent Advances and Future Direction*. London: Croom Helm, 133–48.

— (1986) 'Places and votes: the role of location in the creation of political attitudes', *Urban Geography* 7: 103–17.

Katznelson, I. (1981) *City Trenches*. Chicago: University of Chicago Press.

Kirby, A. (1987) 'The local state and urban politics', *Urban Geography* 8: 273–9.

Lukes, S. (1975) 'Political ritual and social integration', *Sociology* 9: 289–308.

Magleby, D. B. (1984) *Direct Legislation: Voting on Ballot Propositions in the United States*. Baltimore, Md.: Johns Hopkins Press.

Pinch, S. (1985) *Cities and Services: the Geography of Collective Consumption*. London: Routledge & Kegan Paul, London.

Pratt, G. (1986) 'Against reductionism: the relations of consumption as a mode of social structuration', *International Journal of Urban and Regional Research* 10: 377–400.

Saunders, P. (1984) 'Beyond housing classes: the sociological significance of private property rights in the means of consumption', *International Journal of Urban and Regional Research* 8: 202–27.

Staeheli, L. and Hodge, D. (1988) 'The Legimitating Role of Referenda Elections', paper presented at the annual meeting of the Association of American Geographers, Phoenix, Ariz.

Chapter twelve

Spatial targeting strategies, representation, and local politics

Diane Whalley

There is an implied geography to most aspects of urban management. This geography is manifest both explicitly and implicitly: most obviously, the consequences of decisions made in an urban managerial setting are not aspatial – for example, differential tax burdens, the location of noxious facilities and targeted development strategies all involve locational decisions; at a more abstract level, the decision-making process itself is also set within a spatial context. There are over 19,000 different municipal governments in the United States alone, each of which potentially carries its own variant of a management structure, and its own system of representation. The nature of this structure will inevitably influence policy formation and administration. Therefore, since different forms of urban government allow for varying degrees of geographical representation, the form of the municipal electoral process is likely to influence the development and administration of local public policies.

The primary purpose of this chapter is to explore the nature of the relationship between municipal government structure and local policy administration. Representation in local governments has been categorized as falling into five general areas: policy responsiveness; service responsiveness; allocational responsiveness; symbolic representation; and representational focus (Eulau and Karps, 1977; Welch and Bledsoe, 1988). Policy responsiveness deals with the appropriateness of management strategies; both service and allocational responsiveness refer to the provision of services and other benefits, by location. Symbolic representation and representational focus deal with the overall image of representation which is provided. This research is primarily concerned with the impacts of government structure on allocational and service provision issues.

The relationship between government structure and policy administration will be explored through the use of a case study. Elements of the case study include both the selection of a specific municipality and the identification of a relevant policy issue. The most important differences in municipal structure involve differences between the mayor/council and the city manager forms of government, and between at-large and by-ward electoral systems. The case study used in this chapter is based on an analysis of Minneapolis, a city with a mayor/council government structure and a by-ward, partisan electoral system. The policy issue which has been chosen is one which has explicit spatial consequences, in terms of the geographic allocation of scarce funds for housing improvement programmes. The example chosen concerns the administration of Community Development Block Grant (CDBG) funded housing rehabilitation programmes in Minneapolis during 1977. The CDBG programme represents a case in which policies are broadly defined at the federal level but administered locally. Thus these programmes are likely to be applied unevenly both within and across municipalities. They also involve an explicitly spatial targeting process, which forms the basis of the analysis presented here. The chapter will outline the process leading to the designation of a set of neighbourhood target areas, earmarked for housing improvement funds, by examining the transition through the political process from the federal to the local level, and investigating the role of the political structure in Minneapolis as it affected target area delimitation.

In addition, Minneapolis is a city which is characterized by high levels of geographical representation, and also strong neighbourhood activism. It will be argued in this chapter that there is, in fact, a strong correlation between the system of representation and neighbourhood participation in community land use issues. Thus a secondary purpose of the research is to consider the relationship between municipal electoral structure and the politicization of community issues.

Spatial targeting strategies for housing improvement

The 1976 Community Development Act and the 1977 amendments to that Act provided for the inclusion of a set of geographical criteria, which were to be included as a component of federally funded housing improvement schemes. Specifically, it was mandated that assistance provided through the CDBG programme was to be channelled to pre-designated target areas, identified on the basis of resident income and housing conditions. These areas, termed neighbourhood strategy areas (NSAs), were to be selected at the discretion of the local municipalities, with the aim of targeting especially blighted areas and thus concentrating housing improvement efforts in a few areas of extreme need. The programmes available ranged in scope and character from small emergency grants (up to $3,500) to major rental repair. In Minneapolis the rental rehabilitation programmes were hardly used at all, partly owing to a complex set of qualifying criteria, and partly through a lack of incentive for landlords to use them. In practice, the mainstay of the rehabilitation effort in the NSAs was a series of small programmes aimed at exterior maintenance and repairs. The programmes were operated as grants or forgivable incentive loans, pro-rated by income.

There are several advantages associated with neighbourhood targeting strategies. Research has shown that (1) targeting particular neighbourhoods tends to alleviate uncertainty in those areas, and therefore encourages investment (Wilson, 1987); (2) there are significant, but highly localized, economic spill-overs associated with housing rehabilitation strategies (McConney, 1985; Whalley, 1988); (3) the socio-demographic characteristics of a neighbourhood are important in determining the long-term success of rehabilitation programmes (Yap *et al.*, 1978; Mayer and Lee, 1981); (4) other land use dynamics in the neighbourhood will impact on housing strategies (Wolch and Gabriel, 1981; Schlay, 1988); and (5) patterns of access to information and the degree of community participation affect the potential to achieve housing improvement (Smith, 1985). All this suggests that there are distinct geographical effects associated with housing improvement, and that careful targeting of select areas is likely to be beneficial. Target area status is also an advantage from the perspective of the neighbourhood residents. A location within the bounds of a strategy area carries with it access to a set of publicly subsidized housing improvement schemes. Such a location also provides an environment in which private investment may be more likely to occur. This has a long-term impact in terms of both neighbourhood stability and the maintenance of property values. Additionally, neighbourhoods which are defined as target areas often accrue additional advantages deriving from an increased sense of community. These may range from direct advantages such as funding for community facilities to those indirect advantages provided in neighbourhoods with a tightly knit social structure (Abrams, 1984).

Despite the importance of target area delimitation, the criteria for selection are not well defined. Federal guidelines are minimal, and based solely on resident income. The task of NSA designation rests with the individual municipalities. The nature of the selection procedures and criteria chosen is dependent on the existing structure of the municipal government. There is no standardized system of municipal government in the United States, meaning that each of the more than 19,000 urban governments potentially has a different electoral and management structure. The geographic implications of this for target area designation will briefly be outlined, and then the targeting procedures in Minneapolis will be examined in detail.

Government structure

Three major forms of government structure have been identified (Figs. 12.1a–c). In the first two of these general models – the commission form of government and the council/manager form of government – elections are generally non-partisan and city-wide. There is therefore no spatial component to representation in metropolitan governance. In contrast, the mayor/council form of government does provide the potential for geographic representation through the use of a ward system in elections, although not all municipalities utilize the ward system to the same degree. Combinations of by-ward and at-large systems are common for both mayoral and council elections. Further, the relative roles of the mayor and council in municipal governance may vary (Figs. 12.1d–e).

Analyses of various impacts of government structure on municipal management have been inconclusive. Morgan and Pelissero (1980) did not find any significant differences in fiscal efficiency between mayor/council and city manager forms of government. On the other hand, Morgan and Brudney (1985) found that lower spending patterns were weakly associated with 'reformed' cities (i.e. those with at-large, non-partisan elections and professional managers). This evidence is in contradiction with the findings of Lineberry and Fowler (1967) and Clark (1968) who determined that higher spending patterns were associated with reformed cities.

If the impacts of the structure of urban government on management policies are unclear, there are also mixed results concerning the impact of at-large v. by-ward electoral systems. The literature does seem to suggest that minorities are better represented in government in cities with by ward elections (Davidson and Korbel, 1981; Karnig and Welch, 1979, 1981). Ultimately, however, this does not appear to have had consistent impacts on policy-making (Welch and Bledsoe, 1988). Nevertheless, there are some significant findings concerning differences between at-large and by-ward electoral structures. Principally, members elected by-ward tend to have a stronger sense of 'neighbourhood' as constituency, generally resulting in higher levels of factionalism on city councils. This is particularly marked in non-partisan electoral systems (Welch and Bledsoe, 1988).

The implications of these findings on the spatial targeting process are important. In municipalities with strong representation by ward, neighbourhoods competing for target area status are all likely to be better represented than if public officials are selected by city-wide election. On the other hand, a higher degree of competition between neighbourhoods is also likely to be engendered. When the process is dominated by at-large elections a few key neighbourhoods are more likely to receive favourable treatment.

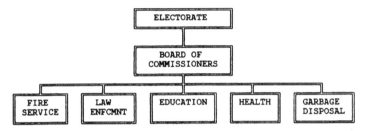

Figure 12.1(a): The commission form of government.
Source: after Short (1984): 207

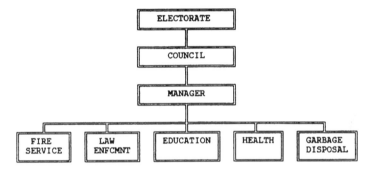

Figure 12.1(b): The council manager form of government.
Source: after Short (1984): 207

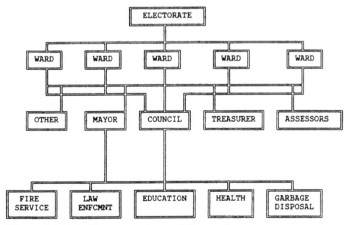

Figure 12.1(c): The mayor/council form of government.
Source: after Short (1984): 207

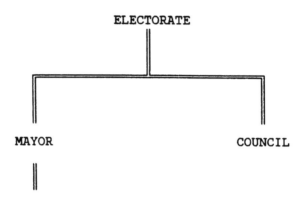

ELECTORATE

MAYOR   COUNCIL

MUNICIPAL DEPARTMENTS

Figure 12.1(d): Weak mayoral system.
Source: after Short (1984): 207

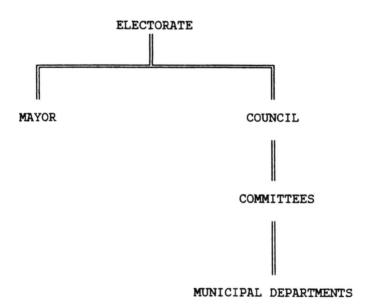

ELECTORATE

MAYOR   COUNCIL

COMMITTEES

MUNICIPAL DEPARTMENTS

Figure 12.1(e): Strong mayoral system
Source: after Short (1984): 207

Given the complexity and variety of municipal government structure, the role of the electoral system in the target area delimitation process is best explored through the use of an example. Issues raised by this study will point to some of the critical components in the process, and will identify areas of future research.

Government structure in Minneapolis

Minneapolis has had an active CDBG target area programme since 1973 and the process of target area delimitation for NSAs was conducted in late 1977 and early 1978. The first CDBG funding was awarded to residents of neighbourhood strategy areas in mid-1979.

The government structure in Minneapolis has incorporated a significant spatial component. The administrative system is a mayoral/council form of government, with each of the thirteen city council members being elected by ward. The mayor is elected at large. In 1978 all terms were for two years. Fig. 12.2 shows the governmental structure in 1978, when the NSA delimitation process was being completed. This structure consisted of a strong mayoral system in which the mayor's office controlled, among other units, the city Planning Department (which was transferred from the city council to the mayor in January 1978), the Minneapolis Housing and Redevelopment Authority (MHRA), and a ten-member Planning Commission. Each of these organizations was made up of a mixture of elected and appointed officials. In addition, in 1978 an eleventh planning district was added to the ten districts which were set up in 1959. In 1979 the Planning District Citizen Advisory Committees (PDCACs), introduced in 1975 to provide citizen input to each of the planning districts, were restructured better to meet affirmative action criteria. As a result, in comparison with many metropolitan areas there is a strong geographical component to both representation and city planning in Minneapolis.

Neighbourhood strategy area delimitation

The task of strategy area delimitation was divided into two stages; the *research* stage was conducted by the city Planning Department, which defined a set of eligible areas. The actual *designations* were determined by the mayor and council. MHRA was then made responsible for the day-to-day administration of the housing improvement programmes.

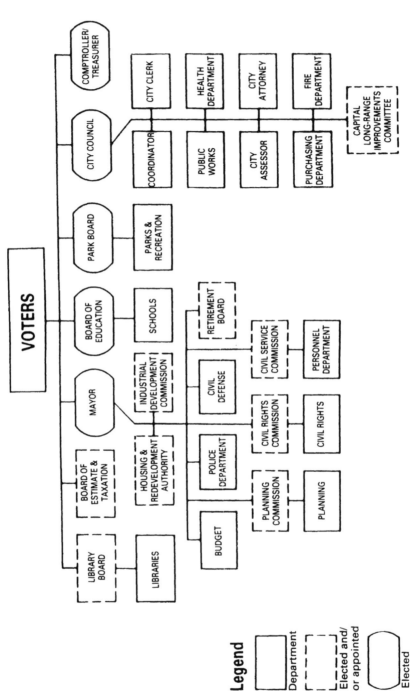

Legend

☐ Department

┌ ┐ Elected and/
└ ┘ or appointed

⬭ Elected

Figure 12.2: The structure of government in Minneapolis

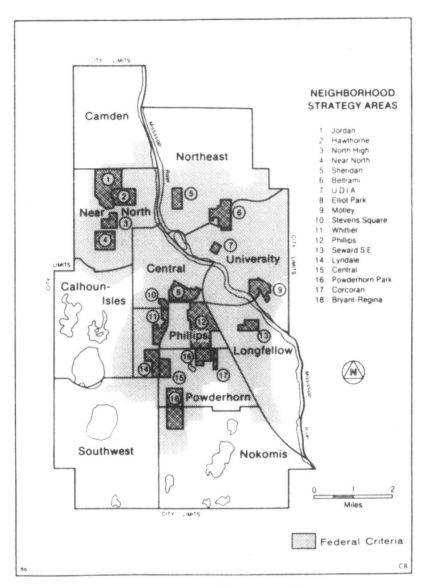

Figure 12.3: Federally defined areas of funding eligibility

The research stage

The criteria which were used in target area delimitation by the Planning Department were based first on the federally defined guidelines concerning resident income, and second on the Planning Commission's own criteria based on housing condition. The federal guidelines were

199

operationalized by identifying areas in which the majority of residents were of 'low' or 'moderate' income. These areas were defined as areas in which median household incomes were below 80 per cent of the metropolitan area median income. At that time the most reliable income information available to the decision-makers was the census tract data from the 1970 census. This information was supplemented using a sample of 1975 income tax returns, which appeared to indicate an income distribution pattern which was roughly consistent with the 1970 pattern. Fig. 12.3 indicates the areas eligible for NSA status, based on the federally defined income criterion and the census and income tax data. The result is a rather broad swathe cut through the middle of the city: an area too large to produce the advantages of selective concentration of funds in smaller, manageable target areas.

A second criterion, one imposed by the Planning Department itself, was based on housing condition. Of particular concern was whether or not the structure conformed to building code standards. Failure to meet code resulted in a designation which categorized the building as being in 'sub-standard physical condition'. Therefore in considering housing condition as a guideline for target area delimitation, the Planning Department imposed the additional qualification that, in order to be eligible for NSA status, an area must be one in which over 25 per cent of the one-and two-family unit structures not previously rehabilitated by public action were classified as 'sub-standard'.

Housing condition was determined from tax assessment records. The city tax assessor's office maintained a classification of all housing units in the city, based on a five-point rating code. The scale ranged from 1 (excellent) to 5 (poor) and took into account existing conditions and the extent of deferred maintenance. The result was that category 4 (fair) described housing with no major structural damage but in need of substantial maintenance investment for purposes such as rewiring, plumbing, weatherproofing, etc. Category 5 represented housing suffering damage to major structural items such as the roof or foundations. It should be noted that the particular loan and grant programmes available were geared to minor, rather than major, repair requirements and thus, in practical terms, were aimed at category 4 housing. Nevertheless, both categories 4 and 5 represented housing which did not meet building code requirements and thus both describe 'sub-standard housing'. This classification therefore formed the basis by which the Planning Department defined eligibility for programme participation. The resultant eligible areas are shown in Fig. 12.4.

All these areas fall within the bounds of the federally eligible areas. The maps presented in Figs. 12.3 and 12.4 were passed from the Planning Department to the mayor and city council for final delimitation of NSAs. The actual NSAs which were defined are marked on both Fig.

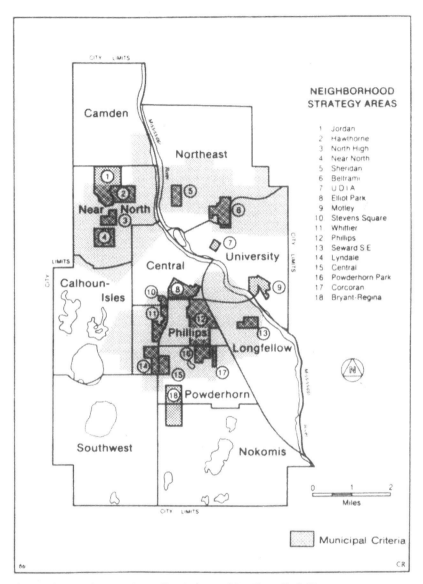

NEIGHBORHOOD
STRATEGY AREAS

1 Jordan
2 Hawthorne
3 North High
4 Near North
5 Sheridan
6 Beltrami
7 U D I A
8 Elliot Park
9 Motley
10 Stevens Square
11 Whittier
12 Phillips
13 Seward S E
14 Lyndale
15 Central
16 Powderhorn Park
17 Corcoran
18 Bryant-Regina

Municipal Criteria

Figure 12.4: Municipally defined areas of funding eligibility

12.3 and Fig. 12.4. An examination of these figures shows that while all of the NSAs fall within the eligible area as defined by federal criteria, several fall outside the Planning Commission's recommended areas, which were based on the distribution of housing condition. The reasons for this are to be found in the political nature of the delimitation process.

Although there is no direct documentation of how the council proceeded from the eligible areas to the actual designations, there is a great deal of supplementary evidence in the form of written reports, memoranda, council meeting minutes and newspaper reports, which lends insight into the process.

The delimitation (decision-making) stage

The stage for the 1977–8 delimitation of NSAs in Minneapolis was actually set several years earlier. Experimental target areas were identified in 1973. These took one of two forms: (1) concentrated improvement projects (CIPs) which were designated for CDBG funding; and (2) neighbourhood emphasis programmes (NEPs) which were targets for the city's own community development programmes. Housing improvement activity in each of these areas was closely monitored by the Planning Department, and appears to have played a later role in influencing NSA status determination. In a memorandum to the city council on 20 April 1978 the mayor expressed a desire to target three types of housing for special improvement funding: (1) seriously blighted housing, in obvious need of repair; (2) transitional housing, which would be likely to benefit from increased maintenance levels; and (3) spot renovation of deteriorated housing in otherwise sound neighbourhoods, to remove blight before it became problematic (City of Minneapolis, 1978: 52).

These three types of housing provide an almost complete geographical coverage of the city of Minneapolis, and extend the eligible target areas out to the federally defined boundaries. While the reasons for this can only be surmised, it should be noted that in a system of municipal government in which both mayor and council are elected to two-year terms, every other year is an election year. This, combined with the ward system, means that community-based issues have tended to be highly politicized in Minneapolis.

Community involvement in the planning process can be observed to have taken place both formally and informally. Through the creation of the PDCACs, neighbourhoods were provided with a formal advisory role. Each of the Minneapolis neighbourhoods has supported its own community newspaper, and analysis of a sample of the issues reported in those newspapers confirms the high level of citizen interest in the planning process. Random selection of press clippings from community newspapers during the 1978-9 period reveals numerous articles concerning land use change issues, and questions of public service provision at the neighbourhood level. Further, members of the municipal governmental structure have themselves used this medium to reach the electorate. Fig. 12.5 serves as an example. In this example, members of the Minneapolis police force campaigned against a fellow officer and would-be council member by running an advertisement in the *Northeaster*, the community newspaper of North-east Minneapolis, one of the neighbourhoods in the 13th council ward.

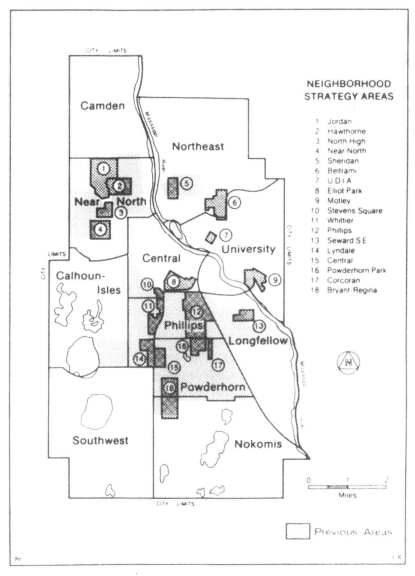

Figure 12.5 Eligible areas under previous programmes

There is thus strong evidence of a territorially based political structure in Minneapolis, combined with heightened levels of community awareness of, and involvement in, the urban political process. This political territoriality serves as a motive for an expanded geographic coverage of target area eligibility, and is consistent with a high degree of inter-neighbourhood competition.

The actual neighbourhoods designated as target areas are well dispersed throughout the eligible area, again representing broad geographical coverage. The initial recommendations of the NSA programme called for twelve designated neighbourhoods; in the final analysis eighteen were targeted, providing each council ward containing eligible housing with at least one NSA (Fig. 12.4). For the most part, the newly-designated NSAs matched the distribution of the existing CIPs and NEPs. In cases where there has been some divergence, either new neighbourhoods have been added, as in the case of the east side university neighbourhoods, or planning reports reveal that the programmes have achieved significant results and have been concluded in the existing project areas, and the targets have simply moved to adjacent areas, as in the north-western neighbourhoods. Overall, the designations reveal a high degree of geographical compromise.

Summary and conclusion

This chapter has attempted to examine some of the relationships between municipal electoral/government structure and the administration of an inherently geographical public programme. Through the exploration of one particular case study, concerning the designation of target areas to be eligible for housing improvement funds, it has been shown that the relationship between these targeting strategies and municipal government structure is potentially an important one. In the case study selected, a system of by-ward council elections every other year has served to make the neighbourhood unit an almost continuous focus of political campaigning. The process of spatial targeting for the allocation of housing improvement funding showed a tendency toward broad, rather than narrow, territorial coverage, resulting from a geographically broad system of representation. Further, as long as the neighbourhood is a consistent focus of political attention there is a potentially strong link between electoral structure and the politicization of community issues.

The variety and complexity of urban governmental structures suggests that many more case studies are necessary to supplement our understanding of the relationships addressed in this chapter. At the same time, the findings presented here suggest that a move toward building more systematic theories of municipal political structure, spatially oriented public policies and community land use concerns, would be a fruitful direction for urban political geographers to pursue.

References

Abrams, P. (1984) 'Realities of neighbourhood care: the interaction between salutory, voluntary and informal social care', *Policy and Politics* 12: 413–29.

City of Minneapolis (1978) *Housing Rehabilitation in Minneapolis, 1973–1977*, Minneapolis: Minneapolis Planning Department.

Clark, T. (1968) 'Community structure, decision making, budget expenditures, and urban renewal in 51 American cities', *American Sociological Review* 33: 576–93.

Davidson, C. and Korbel, G (1981) 'At large elections and minority group representation: a re-examination of historical and contemporary evidence', *Journal of Politics* 43: 982–1005.

Euleau, H. and Karps, P (1977) 'The puzzle of representation: specifying components of responsiveness'. *Legislative Studies Quarterly* 2: 233–54.

Karnig, A. and Welch, S. (1979) 'Sex and ethnic differences in municipal representation, *Social Science Quarterly* 60: 465–81.

—— and S. Welch (1981) *Black Representation and Urban Policy*. Chicago: University of Chicago Press.

Lineberry, R. and Fowler, E. (1967) 'Reformism and public policies in American cities', *American Political Science Review* 61: 701–16.

Mayer, N. and Lee, O (1981) 'Federal repair programs and elderly homeowners' needs', *Gerontologist* 21: 312–22.

McConney, M. (1985) 'An empirical look at housing rehabilitation as a spatial process', *Urban Studies* 22: 39–48.

Morgan, D. and Brudney, J (1985) 'Urban Policy and the City Government Structure: Testing and Mediating Effects of Reform', paper presented to the annual meeting of the American Political Science Association, New Orleans.

—— and J. Pelissero (1980) 'Urban policy: does political structure matter?' *American Political Science Review* 74: 999–1006.

Schlay, A. (1988) 'Not in that neighbourhood: the effects of population and housing on the distribution of mortgage finance within the Chicago SMSA' *Social Science Research* 17: 137–63.

Short, J. (1984) *An Introduction to Urban Geography*. London: Routledge & Kegan Paul, London.

Smith, R. (1985) 'Activism and social status as determinants of neighbourhood identity', *The Professional Geographer* 37: 421–32.

Welch, S. and Bledsoe, T. (1988) *Urban Reform and its Consequences: a Study in Representation*. Chicago: University of Chicago Press.

Whalley, D. (1988) 'Neighborhood variations in moderate housing rehabilitation program impacts: an accounting model of housing quality change', *Economic Geography* 64: 45–61.

Wilson, D. (1987) 'Urban revitalization on the upper West Side of Manhattan: an urban managerialist assessment', *Economic Geography* 63: 35–47.

Wolch, J. and Gabriel, S. (1981) 'Local land development policies and urban housing values' *Environment Planning A* 13: 1153–72.

Yap, L., Greenston, P., and Sadacca, R. (1978) *Lower Income Housing Assistance Program (Section 8): Nationwide Evaluation of the Existing Housing Program*, Washington, D.C.: US Department of Housing and Urban Development, Office of Policy Development and Research.

Chapter thirteen

Identical geography, different party: a natural experiment on the magnitude of party differences in the US Senate, 1960–84

Bernard Grofman, Robert Griffin and Amihai Glazer

Inspired by Downs (1957), dozens of articles have been written in which party competition is modelled as a battle by office-seeking politicians for the allegiance of the median voter (see reviews in Riker and Ordeshook, 1973; Enelow and Hinich, 1984). Yet the critical implications which derive from this median voter mode, namely Tweedledum and Tweedledee politics and politically competitive elections, are falsified by the data.

The empirical evidence for the United States contradicts Downs's implications in several ways. (1) Platforms of opposing political parties are far from identical (Page, 1981; Pomper, 1969), and despite candidate penchants for generality and ambiguity, recent US presidential candidates have often shown striking contrasts in candidate positions. (2) When congressional districts change hands the new House members vote differently from the old if there has been a change in party, but not otherwise (Brady and Lynn, 1973; Clausen, 1973; Fiorina, 1974; Glazer and Robbins, 1983). (3) Voting differences between senators of opposite parties persist even after constituency characteristics have been controlled for (Bullock and Brady, 1983; Poole and Rosenthal, 1985). (4) Most incumbents are re-elected (Mayhew, 1977). (5) Several areas of the country have remained for long periods of time under one-party control.

We make use of a 'natural experiment', comparisons of the voting records of senators from the same state but of opposite parties. This allows us to investigate the magnitude of the differences in ideology caused by party differences when the potentially confounding effects of constituency differences have been completely controlled for. This use of a natural experiment distinguishes our analysis from that of earlier work (e.g. Bullock and Brady, 1983) on the Senate. Our key findings about the US Senate fly in the face of a simple Downsian model:

1. US senators from the same state but of opposite parties vote quite distinctly from one another, while differences between senators from the same party and the same state are minimal. Moreover, in states where

one senator is a Democrat and the other a Republican, the Democrat is almost always to the left of the Republican from that state.

2. For most of the past 25 years the mean ideological difference between senators of the same state who are of opposite parties exceeds that between two randomly chosen senators of opposite parties in the nation as a whole.

Competition for the US Senate, 1960–84

A number of measures of Congressional liberalism are based on roll calls, of which the best known are those issued by the Americans for Democratic Action (ADA) and by the Americans for Constitutional Action (ACA). In addition, political scientists (Manley, 1981; Ornstein *et al.*, 1984) make use of the Conservative Coalition (CC support score) which indicates the proportion of issues on which members vote with conservative southern Democrats. Kritzer (1978: 492) shows that these measures of liberalism are almost perfectly correlated with each other (as well as with the roll-call scores produced by other organizations) and essentially tap a single dimension (see also Kau and Rubin, 1982; Poole and Rosenthal, 1985). In 1981, for example, the correlation between ADA score and CC score was -0.94 in the Senate and -0.93 in the House. To simply our exposition, we deal exclusively with ADA scores.

We use ADA score differences between Republican and Democratic senators for the years 1960–84 to determine the degree to which Democrats differ from Republicans when geographic constituency is held constant. The standard interpretation of Downs (e.g. Davis *et al.*, 1970) would predict that senators from both parties are to be found at (or very near) the overall median voter in the state. However, *senators from the same state differ if and only if they are of opposite parties, and Democrats are virtually always ideologically to the left of the Republicans* (see Table 13.1).

As shown in Table 13.1 for the period 1960–84, the average absolute difference between the ideologies of Democratic and Republican senators from the same state is a huge 43.4 points. Furthermore, the general pattern holds up every year, even in the 1980s when a much higher percentage of divided party states are in the south. Moreover, other evidence shows that in states with senators of opposite parties 93 per cent of the time the Republican senator was to the right of the Democrat from that state. The differences can be quite extreme. For example, in New York in 1982 Daniel Moynihan, a Democrat, had an ADA score of 95 while Alfonse D'Amato, a Republican, had an ADA score of only 15.

Table 13.1: Mean absolute difference in ADA score: party composition, Senate, 1960–84. (N in brackets)

Year	D/D states	D/R states	R/R states
1960	12.8 (24)	52.1(15)	18.6(9)
1961	14.2 (24)	64.0 (15)	20.0 (7)
1963–4	14.8 (25)	59.1 (16)	13.9 (8)
1965	15.5 (26)	53.1 (16)	12.5 (8)
1966	16.0 (24)	51.8 (17)	16.9 (8)
1967	18.8 (22)	44.4 (18)	12.7 (9)
1968	20.2 (22)	32.5 (17)	24.8 (9)
1969	17.4 (17)	49.3 (21)	17.1 (11)
1970	19.5 (17)	49.8 (22)	14.6 (10)
1971	13.8 (17)	52.9 (18)	19.6 (13)
1972	13.2 (17)	41.9 (18)	20.0 (13)
1973	11.1 (18)	43.7 (20)	23.2 (11)
1974	8.1 (17)	37.8 (22)	24.3 (10)
1975	10.0 (21)	31.7 (19)	24.6 (9)
1976	19.5 (21)	32.1 (19)	27.8 (9)
1977	14.7 (19)	33.9 (23)	19.3 (7)
1978	15.7 (19)	33.3 (23)	11.4 (7)
1979	13.4 (16)	31.0 (26)	15.9 (7)
1980	18.4 (16)	30.5 (26)	11.7 (7)
1981	13.5 (11)	44.4 (24)	9.6 (14)
1982	18.0 (10)	42.7 (25)	10.0 (14)
1983	12.7 (11)	40.4 (23)	12.0 (15)
1984	17.3 (11)	45.4 (23)	10.6 (16)
Average 1960–84	15.2	43.4	17.0

The party differences shown in Table 13.1 are not affected by the 'south is different' phenomenon to any significant extent. In the south the average ADA difference between senators of opposite party from the same state is 36.2. Of course, there have been only a handful of southern senators from divided states. Table 13.2 compares senators from the same state who held office at the same time. To see whether the party-effect phenomenon holds over time, we track the ADA scores of each state's senators over time, as incumbents change and as the party of the incumbent shifts.

As Table 13.2 shows, the effect is present with both longitudinal and cross-sectional data, and with the same dramatic differences between succeeding senators of different parties as we observed in our cross-sectional, within-state, comparisons. We see that, on average, the ADA score of a Democratic senator is 32.7 points higher than that of the Republican senator he replaced; partisan shifts in the reverse direction are associated with an average decrease of 52.6 points in ADA scores. In contrast, individual senators change very little over time, and shifts in Senate seats not associated with a change in parties have only a minimal effect on ADA ratings. For example, in Iowa, when the Republican, Hickenlooper, was replaced by the Democrat, Hughes, in 1969, the shift in ADA score was a whopping 89 points, from 0 to 89; when Hughes was replaced by Culver, another Democrat, in 1975, the shift was only 14 points, from 86 to 100; but when the Democrat, Culver, was replaced by a Republican, Grassley, the shift was again dramatic, from 78 to 5. Similarly, in 1961 when the Republican, Martin, was replaced by Miller, another Republican, the shift, although larger, from 30 to 0, was far smaller than the shift when Miller was replaced by a Democrat, Clark, in 1973. Then the shift in ADA score was 60 points, from 20 to 80; and when the Democrat, Clark, was replaced in 1979 by the Republican, Jepsen, the ADA score dropped massively from 90 to 0. We might also note that the mean value of shifts that did not involve a change in parties was even smaller than the mean absolute value of such shifts reported in Table 13.1; -1.4 points for seats that stayed Democrat and -0.4 for seats that stayed Republican.

Table 13.2: Mean longitudinal shift in Senatorial ADA scores across all states in 1960–1984 by region and by categories of party and incumbent change. (N in brackets)

	Change in mean ADA Scores					
Region	Same person D/D	Different person D/D	Same person R/R	Different person R/R	Different person R/D	Different person D/R
South	3.7 (56)	6.1 (11)	1.1 (12)	0.2 (4)	19.2 (5)	-27.8 (13)
Mid West	1.8 (37)	7.3 (6)	-0.6 (25)	5.7 (6)	60.6 (10)	-40.5 (12)
North-east	3.4 (33)	5.7 (3)	-7.4 (31)	-2.7 (7)	49.2 (13)	-11.2 (9)
West	2.5 (40)	2.3 (6)	-2.3 (26)	3.4 (7)	68.9 (8)	-42.8 (16)
All states	2.9 (166)	5.5 (26)	-3.1 (94)	1.7 (24)	52.6 (36)	-32.7 (50)

There are some minor regional differences. In the Mid West, for example, a shift from a Democratic to Republican Senator led to a 40 point drop in ADA score, while in the north-east the drop was only 11 points. In the Mid West, a switch from a Republican to a Democrat generated a 60 point increase in ADA score; in the north-east the shift was a slightly less sizeable 48 points. On balance, party differences were least in the south. In all regions, maintenance of the same party, but with a different incumbent, on average, involved less than a 10 point shift in ADA score; while if we tracked the same senator over time his or her score shifted on balance by a minuscule amount, usually well under 5 points, with a very slight tendency for Democrats to become more liberal and Republicans to become more conservative. This latter effect, however, may well be a simple regression to the mean phenomenon. In short, party really matters even when geographic constituency is held constant.

Our analysis reveals another important and rather unexpected finding: *during the period 1960–84 the mean ADA difference between senators from the same state exceeded that between randomly chosen senators from the country as a whole.* Table 13.3 shows mean Republican and Democratic ADA scores for the senate as a whole. For the period 1960–84 the mean ADA difference between Republican and Democratic senators was only 34.0, compared to a mean absolute difference of 43.4 for senators from the same state of opposite parties (and a mean difference of 42.5). Comparing the rows in Tables 13.1 and 13.3 we find that this pattern holds strongly from 1960 to 1974, while within-state and between-state differences are virtually identical through 1980; since 1980 the pattern has reversed, with the between-state differences between parties greater than the within-state ones.

Why are between-state ADA differences between senators of opposite parties now greater than within-state differences, although in previous decades the difference was in the other direction? We believe the best explanation is a compositional one. The simple answer is the steady growth of Republican senatorial strength in the south, and the diminishing ideological heterogeneity of the Democratic Party in the Senate, as shown by the long-run decrease in the standard deviation of Democratic Party ADA scores.

The standard deviation of Democratic ADA scores has fallen by roughly one-third from the 1960s to the 1980s (see Table 13.3). This fall is almost perfectly paralleled by an almost perfectly monotonic decline in Democratic senators from the thirteen states of the old south: from 24 in 1960 to 14 in 1984. The loss of these conservative senators on the Democratic side of the aisle and their replacement by even more conservative Republicans on the Republican side of the aisle has led to an increase in the liberalism of Senate Democrats relative to Senate Republicans.

Table 13.3: Mean and standard deviation of ADA scores for Republican and Democrat Senators 1960–84

Year	Republicans			Democrats			
	N	Mean	SD	N	Mean	SD	Partisan difference
1960	34	30.0	25.7	64	61.8	37.9	31.8
1961	32	26.2	28.0	64	70.7	33.7	44.5
1963–4	33	28.9	24.9	66	60.4	32.8	31.5
1965	32	20.8	26.6	68	63.1	33.2	42.3
1966	33	20.4	26.9	66	59.1	34.1	38.7
1967	36	24.6	24.7	63	52.4	29.5	27.8
1968	36	30.5	28.6	62	43.1	30.6	12.6
1969	43	33.2	31.3	56	61.8	33.4	28.6
1970	42	29.0	29.4	57	56.5	32.0	27.5
1971	44	25.1	24.3	54	64.8	29.2	39.7
1972	44	24.8	24.4	54	50.6	29.8	25.8
1973	43	26.4	26.5	56	62.8	27.4	36.4
1974	43	31.0	29.2	56	61.6	29.8	30.6
1975	38	31.4	30.2	61	61.5	30.1	30.1
1976	38	25.7	29.2	61	55.1	25.9	29.4
1977	38	25.1	27.4	61	58.9	25.9	33.8
1978	38	25.9	22.1	61	52.9	22.8	27.0
1979	41	22.8	19.7	58	49.1	23.0	26.3
1980	41	29.3	19.5	58	58.7	19.8	29.4
1981	53	16.8	16.0	45	66.2	25.6	49.4
1982	54	25.1	23.6	45	68.7	23.9	43.6
1983	54	22.4	19.4	45	68.3	21.7	45.9
1984	55	24.7	23.7	45	73.9	23.4	49.2
Average		26.1			60.1		34.0

Discussion

In looking for explanations for the importance of party in generating differences in ideological voting behaviour among senators and congressmen elected from the same state but of opposite parties, one basic intuition is that political competition in the US has both a centralizing and a decentralizing component. The centralizing component is the widely shared view of a difference between Republicans and Democrats, with the latter seen as being to the left of the former. The decentralizing component is the fact that national party competition must take place in 50 states (or 435 Congressional districts) which are distinct constituencies, each with very different attitudinal and demographic characteristics. Because of the interaction of national and local effects, diversity at the constituency level is reconciled with similarities at the national level, even though some Democrats may look a lot like Republicans (and vice versa) if judged relative to the *national* mean of each party (see especially Weinbaum and Judd, 1970: 300–1 on the national party effect, and theoretical work on the behaviour of national parties faced with multiple constituencies, Austen-Smith, 1983, 1984, 1986).

Another key intuition is the recognition that a candidate's election constituency may be very different in character from his/her geographic constituency. Fenno (1978) calls attention to the 'concentric circles' of contact between representatives and their constituencies, with a few individuals playing a critical role in campaigns and others having direct access to the candidate when they seek it. Fenno (1978) also emphasizes a candidate's 'election constituency', the set of individuals who actually voted for him/her. Markus (1974) and Bullock and Brady (1983) show that the characteristics of the election constituency are far more important in predicting a representative's vote than the characteristics of the overall geographic constituency which the representative ostensibly serves. Similarly, other critics of Downs (such as Hirschman, 1970) point out that the Downsian model is inappropriate if the more extreme wings of each party gain control over party policy and eschew the middle ground. Moreover, as many sociologists and political scientists (including some who are sympathetic to a rational choice approach) continue to stress, there are group bases of political competition (see, e.g., Fiorina, 1974; Axelrod, 1972, and updates in 1976, 1980 and 1984; cf. Froman, 1963). The same point is often made by practitioners of what has been called the 'new political history' (Benson, 1961; Jensen, 1971; Kleppner, 1970, 1979; Silbey, 1985; cf. Burnham, 1970; McCormick, 1974).

In viewing the 19th century, most contemporary historians have decisively rejected the Downsian view of parties in favour of a group-rooted basis of partisan conflict (Silbey, 1985: 59). Customarily, 20th century party politics are viewed as more nearly fitting a Downsian model. However, Silbey's (1985: 59–61) language about antebellum party politics may apply, perhaps even with undiminished force, to contemporary party politics as well

> The stress on parties as primarily political machines, while appropriate on one level, does not confront the other dimensions present nearly enough ... Elite maneuvering for partisan advantage always went on. But it did not occur within an anything goes for electoral victory mentality. It could not ... Different social groups were attracted to each party by each's stance and where their own friends and enemies were located themselves. Parties had

different centers of gravity based on their component groups. There were clearly distinctive political mind sets in America. There was a Democratic mentality and a Whig, later Republican, mentality. Neither party could digest every demand, interest and pressure within society at a given moment. They could never be all things to all men.

At least in the US Senate it is apparent that party does matter, and party and election constituency effects are stronger than geographic constituency ones. The effect of party constituencies is reinforced by the importance of policy-motivated activists as campaign workers and contributors and by the role of internal selection mechanisms such as party primaries in constraining the policy positions of party nominees (see especially Aldrich, 1980; Coleman, 1972; Aranson and Ordeshook, 1972).

We believe it useful to rethink the implication of much of the previous empirical work on the link between representative and constituency. For example if the constituency link is to a representative's election constituency and not the geographic constituency, then classic findings such as those of Miller and Stokes (1963) or Cnudde and McCrone (1966) may understate (or mistake) the extent to which a representative's votes are the product of potential electoral sanctions. While some authors (e.g. Markus, 1974; Fenno, 1978; Bullock and Brady, 1983; and McCubbins and Sullivan, 1984) have understood this point, the Downsian perspective continues to lead most political scientists to search for relationships between a representative's views and those of the *geographic* constituency.

At least as judged by our experience in relating our findings to other scholars, even political scientists who understand the difference between electoral and geographic constituency underestimate the remarkable magnitude of the direct party effect, and attribute most of the difference between senators of different parties to compositional differences in geographic constituency, i.e. to the fact that Democrats (except in the south) tend to be elected from more liberal states and thus are more liberal. Yet the multivariate correlation between a host of economic and demographic variables (including race, income, and employment) and Senate ADA scores in 1982 was just 0.54 and rises to only 0.58 even with a dummy variable for the south. However, when we include party share of the 1980 presidential vote in the state the correlation rises to 0.83. Indeed the simple bivariate correlation in 1982 between party vote share and ADA score is 0.68 for the Senate as a whole; and it is even higher if we control for section. In particular, the simple correlation between party vote share and ADA score is 0.74 for non-Southern senators and 0.83 for southern senators – an effect *stronger than for all demographic variables combined.* Moreover, even if we try to predict the *mean* ADA score of the senators from each state, demographic variables together with region account for only 47 per cent of the variance in 1982. However, if we introduce a further dummy for party control (1 if 2D, -1 if 2R), the r^2 increases to 0.67.

While there is a compositional effect, which affects which states will elect Democrats, geographic constituency cannot explain why senators of *opposite parties* of the *same state* are so very different in their ideology. *We cannot understand politics in the United States in terms of the simple Downsian model which posits convergence to a unidimensional median.* We must introduce complicating institutional factors such as competition across multiple constituencies (Austin-Smith, 1984, 1986) and internal party selection mechanisms (Aranson and Ordeshook, 1972; Coleman, 1972); as well as mechanisms which tend to create distinct electoral constituencies for each of the parties (Aldrich, 1983; Glazer, 1988). Remarkably, however, multidimensionality in the space of political competition may not be a major confounding factor. Also, one other possible explanation of within-state party differences (adapted from Baron, 1984; cf. Fiorina, 1987), the notion that in picking senators voters try to balance them off so as to achieve a desirable aggregate ideological outcome, was not supported, although our empirical test was crude.

References

Aldrich, J. H. (1980) *Before the Convention. Strategies and Choices in Presidential Nomination Campaigns.* Chicago: University of Chicago Press.
—— (1983) 'A Downsian spatial model with party activism', *American Political Science Review* 77: 974–90.
Aranson, P. and Ordeshook, P. (1972) 'Spatial strategy for sequential elections', in R. Niemi and H. Weisberg (eds.) *Probability Models of Collective Decision Making*, Columbus, Ohio: Merrill.
Austen-Smith, D. (1983) 'The spatial theory of electoral competition: instability, institutions, and information'. *Environment and Planning C: Government and Policy* 1: 439–59.
—— (1984) 'Two-party competition with many constituencies', *Mathematical Social Sciences* 7: 177–98.
—— (1986) 'Legislative coalitions and electoral equilibrium', *Public Choice* 50: 185–210.
Axelrod, R. (1972) 'Where the votes come from: an analysis of electoral coalitions 1952–1968', *American Political Science Review* 66: 11–20 (plus 1976, 1980, and 1984 updates).
Baron, A. (1984) 'The Baron Report', No. 218, 19 November: 2.
Benson, L. (1961) *The Concept of Jacksonian Democracy: New York as a Test Case.* Princeton, N.J.: Princeton University Press.

Brady, D. and Lynn, N. (1973) 'Switched seat congressional districts: their effect on party voting and public policy', *American Journal of Political Science* 67: 523–43.

Bullock, C. and Brady, D. (1983) 'Party, constituency, and roll-call voting in the U.S. Senate', *Legislative Studies Quarterly* 8: 29–43.

Burnham, W. D. (1970) *Critical Elections and the Mainsprings of American Politics*. New York: Oxford University Press.

Clausen, A, (1973) *How Congressmen Decide: a Policy Focus*. New York: St. Martins.

Cnudde, C. F. and McCrone, D. O. (1966) 'The linkage between constituency attitudes and congressional voting behavior: a causal mode', *American Political Science Review* 60: 66–73.

Coleman, J. (1972) 'The positions of political parties in elections', in R. Niemi and H. Weisberg (eds.) *Probability Models of Collective Decision Making*. Columbus, Ohio: Merrill, 332–57.

Davis, O., Hinich, M. C. and Ordeshook, P. C. (1970) 'An expository development of a mathematical model of the electoral process, *American Political Science Review* 64: 426–48.

Downs, A. (1957) *An Economic Theory of Democracy*. New York: Harper and Row.

Enelow, J., and Hinich, M. (1984) *The Spatial Theory of Political Competition: an Introduction*. Cambridge: Cambridge University Press.

Fenno, R. (1978) *Homestyle. House Members and their Districts*. Waltham, Mass.: Little Brown & Co.

Fiorina, M. (1974) *Representatives, Roll Calls, and Constituencies*. Lexington Mass.: Lexington Books.

—— (1987) 'The Reagan years' in B. Cooper and A. Kornberg (eds.) *The Resurgence of Conservatism in Britain, Canada and the U.S.* Durham, N.C.: Duke University Press.

Froman, L. A. (1963) *Congressmen and their Constituents*. Boston, Mass.: Little Brown.

Glazer, A. (1988) 'A Formal Model of Expressive Voting', unpublished manuscript.

—— Owen, G. and Grofman, B. (1989) 'Imperfect information models of party competition', *Mathematical Modelling* (forthcoming)

—— and Robbins, M. (1983) 'Voters and roll-call voting: the effects on congressional elections', *Political Behavior* 6(4): 377–89.

Hirschman, A. O. (1970) *Exit, Voice, and Loyalty*. Cambridge, Mass.: Harvard University Press.

Jensen, R. J. (1971) *The Winning of the Midwest: Social and Political Conflict, 1888–1896*. Chicago: University of Chicago Press.

Kau, J. B. and Rubin, P. (1982) *Congressmen, Constituents, and Contributors: Determinants of Roll Call Voting in the House of Representatives*. Boston, Mass.: Martinus Nijhoff.

Kleppner, P. (1970) *The Cross of Culture: a Social Analysis of Midwestern Politics*. New York: Free Press.

—— (1979) The Third Electoral System, 1853–1892. Chapel Hill, N. C.: University of North Carolina Press

Kritzer, H. M. (1978) 'Ideology and American political elites', *Public*

Opinion Quarterly 42: 484–502.

McCormick, R. L. (1974) 'Ethno-cultural interpretations of nineteenth century American voting behavior', *Political Science Quarterly* 89: 351–77.

McCubbins, M. D. and Sullivan, T. (1984) 'Constituency influence on legislative policy choice', *Quality and Quantity* 18: 299–319.

Manley, J. F. (1981) 'The conservative coalition in Congress', in L. Dodd and B. Oppenheimer (eds.) *Congress Reconsidered* First Edition, Washington, DC.: *Congressional Quarterly*, 75–95.

Markus, G. B. (1974) 'Electoral coalitions and Senate roll-call behavior: an ecological analysis', *American Journal of Political Science* 18(3): 595–605.

Mayhew, D. (1977) *The Electoral Connection*. Newhaven, Conn.: Yale University Press.

Miller, W. E. and Stokes, D. (1963) 'Constituency influence in Congress', *American Political Science Review* 57: 45–56.

Ornstein, N. J., Mann, T. E., Malbin, M. J., Schick, A. and Bibbey, J. F. (1984) *Vital Statistics on Congress 1984–1985 Edition*. Washington, D.C.: American Enterprise Institute for Public Policy Research.

Page, Benjamin (1981) *Choices and Echoes in Presidential Politics*. Chicago: University of Chicago Press.

Pomper, G. (1969) *Elections in America: Control and Influence in Democratic Politics*. New York: Dodd Mead.

Poole, K. and Rosenthal, H. (1985) 'A spatial model for legislative roll-call analysis, *American Journal of Political Science* 29: 357–84.

Riker, W. and Ordeshook, P. (1973) *An Introduction to Positive Political Theory*, Englewood Cliffs, N.J.: Prentice-Hall.

Silbey, J. (1985) *The Partisan Imperative: the Dynamics of American Politics Before the Civil War*. New York: Oxford University Press.

Weinbaum, M. G. and Judd, D. R. (1970) 'In search of a mandated congress', *Midwest Journal of Political Science* 14: 276–302.

Chapter fourteen

Local voting and social change

Nancy Ettlinger

Voting is immediately intelligible as an act of citizenship that provides individuals with the opportunity to voice their views formally on issues and to select representation. Yet the local context of voting is complicated because individual or group voting is influenced by the existing social, economic, and political milieu. Moreover, the political or socio-economic climate can affect whether the outcome of an election is trivial or significant in effecting long-term change. Accordingly, we might raise the question as to whether voting enables individuals or groups to bring about change or create issues, or whether voting re-creates or passively reflects political and socio-economic preconditions. From the perspective of voting as a means of instituting social change, I will discuss the preceding chapters 10, 11, and 12.

In Chapter 10 Roberts *et al.* recognize the ability of people to institute desired changes through the local political process, yet they find that overarching structural conditions limit the outcome and cast a dim shadow over the efforts of the family farmer. The structures of the macro-economy and local production converge to dwarf both local politics and corporate intentions. Macro-economic conditions imperil family farming as increased mechanization, capital intensity, and debt stress result in production systems that take advantage of economies of scale, require larger amounts of land, and are most amenable to large agribusinesses. Accordingly, corporate farming is mandated by structural conditions, irrespective of the agendas of corporate executives. In this regard, populist, anti-corporate sentiment appears economically unsound. However, if we extend Hodge and Staeheli's thesis of urban politics (Chapter 11) to an agrarian setting, then issue-oriented elections are predictably consumption and not production-oriented, and agrarian populist behaviour represents a rational response to conditions that threaten private property. Yet the economic need for economies of scale, specifically in the meat-packing industry, means that local political efforts are inconsistent with the economic requirements of local production. The result: production requirements will overshadow local consumer self-interest in the long run, because farmers mistakenly try to institute change by addressing the effects rather than the structure of production.

The structure of production can probably be addressed most effectively by a unified front of farmers at the scale of the regional economy. However, it is unlikely that farmers in a regional economy can unify, in part because the spatial covariation of productivity and weather patterns means that farmers in different places are affected differently. Thus farmers in a regional economy are too fragmented across space to react uniformly (and beyond the local level) to overarching economic structure, especially in the context of a multiplicity of short-run pressures, such as debt stress. In this regard, farmers as a group may unwittingly act in concert with economic structure to minimize the impact of a successful referendum.

The problems of achieving social change through local voting in an agrarian context do not, however, extend to the urban context. The spatial separation, or lack thereof, between work and home differentiates the agrarian and urban contexts and defines the relationship between production and consumption systems (Katznelson, 1981). The unity of work and home in the family farm means that production requirements and local consumer values must compete because they overlap in the same locale; i.e. the protection of private property (the farm) is the same as the protection of the means of production. In the city, the spatial division of work and home permit relative independence of these two spheres.

In Chapter 11 Hodge and Staeheli focus less on this geographic element and more on aspatial processes in explaining the emergence of an independent consumption dimension in urban life. They contend that a consumption system independent of production depends upon the economic enfranchisement of the populace. Such relative independence of consumer values is, however, historically contingent upon production, for it is the result of the evolution of Fordist production, which was the avenue by which working class *per capita* incomes increased in the United States. This processual perspective, although developed for the urban context, can provide some insight into the plight of the family farmer. As such, the historically contingent conditions in the agrarian context obstruct the eventual independence of the consumption dimension because the changing requirements of agricultural production in the United States economically disenfranchise the family farmer and empower large agribusinesses.

Despite the logical appeal of Hodge and Staeheli's processual approach, I question the conclusion on empirical grounds. First, from the standpoint of voter turnout and the socio-economic composition of those who turn out to vote, Hodge and Staeheli's thesis regarding voting behaviour is a foregone conclusion. Given the strong and positive relationship between affluence and/or education and voter turnout in twentieth century America (Burnham 1987; Hill and Kent, 1988), economic enfranchisement is more a precondition for the active electorate in partisan politics than an explanation for distinct spheres of voting behaviour. As such, whether the context is economic upswing or downswing, Fordist or post-Fordist, we might expect the same voting behaviour because the same economically enfranchised sub-populations represent the majority of active voters. Second, within a capitalist system, and irrespective of economic enfranchisement, we might expect that people will vote as members of a group or party in situations in which they personally are uninvolved (such as voting for a candidate); conversely, we might expect that people will vote in their own self-interest in situations in which they personally are involved (such as voting on local issues). Indeed, this logic may help clarify why Hodge and Staehili's framework allows for Democratic voting in candidate elections and significantly more conservative (Republican-oriented) voting in issue elections, but conceptually cannot permit the converse situation. (Empirically this latter situation may occur, but the conceptual framework cannot explain it.)

Although Hodge and Staeheli do not purport to explain agrarian voting, their theoretical framework should be capable of explaining voting behaviour in different contexts, from a comparative point of view. Given their contention that economic enfranchisement enables the separation of the consumption and production dimensions in voting, we would expect voting behaviour to differ between economically disenfranchised family farmers and economically enfranchised urban residents. Yet family farmers and urban residents alike vote in local issue elections for their own self-interest and to protect their private property. Thus, while Hodge and Staeheli's attention to the distinctness of consumption and production-oriented urban voting profitably informs us, the explanation of this distinctness is problematic and cannot explain successfully variation from one context to another.

I suggest that the difference between the agrarian and urban contexts is not in voting behaviour itself, but in the significance (or lack thereof) of the outcome of the vote in effecting change. In this regard, the inescapable competition between the consumption and production spheres on the family farm, together with farmers' overall fragmentation and concomitant disunified and differentiated behaviour, ensures only a hollow victory in local issue elections; in contrast, the separation of home and work in the city at least permits a more solid victory in local politics.

Although none of the three preceding chapters explicitly addresses the possibility of effectively bringing about positive social changes in the urban context, Whalley's Chapter 12, in particular, is ripe with implications. This case study reveals a blend of people and place-oriented urban policy: i.e. rehabilitation in Minneapolis was targeted towards people of a particular income level as well as towards places (where low income groups live) that were characterized by place-specific problems (e.g. zones of blight, transition, or spot renovation). Interestingly, the final result was wide geographical dispersal of municipal funds, as every neighbourhood received a zone designated as a target area. This impressive geographic coverage of funding was a function of the local political process, and reflected more the distribution of the metropolitan constituency and municipal government structure than the distribution of need. This is not to say that unneedy neighbourhoods received funds; rather, that political, more than socio-economic, factors figured most prominently in determining the distributional outcome of municipal funding.

From the standpoint of social change, it is advisable to raise the question as to whether such political causality is predictable. Welch and Bledsoe's (1988) comparative analysis of urban reform and representation indicates that the impact of political structure on policy is quite small. If we accept this conclusion, then Whalley's case study suggests that political structure may affect policy, depending upon the frequency of challenge to geographic representation. Specifically, if urban political structure is based on a district (rather than an at-large) system, and if geographic representation is frequently challenged (e.g. every other year in the case of Minneapolis), then municipal funding will be evenly distributed. At least, we might consider such a possibility as a hypothesis, amenable to testing across a number of communities. We also might consider whether the converse situation would hold across a number of communities, i.e. would an at-large system and infrequent challenge to geographic representation result in unevenly distributed municipal funding? Such questions can lead us to identify the relationship between urban political systems and the equity or efficiency of municipal policy, facilitating evaluation of the type of political system that may be most desirable in socio-economic terms.

If we consider that a particular local political system is capable of engendering social change, then it behoves us to consider the type of change that may occur (Ettlinger, 1985; Stohr, 1982). In particular, we should be careful not to overstate the social consequences of equity-oriented policy in the case that structural, qualitative change is not achieved; also, we should not understate the economic consequences of efficiency-oriented policy that increases productivity, perhaps at the expense of social change. Although equity and efficiency strategies are often perceived as competitive (Berentsen, 1981), they may share common ground from the standpoint of social change, since neither may result in qualitative, structural change. For example, an equitable distribution of social service funds in an urban area may in the short-run alleviate some stress by providing basic needs, but such policy is typically incapable of altering the overall structural position of the poor (Hirschman, 1981, chapter 1). Moreover, from the standpoint of the theory of social control, the provision of social services and associated funds to those in need may be a means by which to placate potentially violent or disturbing factions of the lower class. For example, Erie (1985) has noted that black voter participation rates sharply declined during the years of welfare-state expansion, from the mid-1960s to the late 1970s. What, then, are the possibilities for positive, local, long-term, structural change, and can voting make a difference?

Answering such a complex question may be a far-sighted goal; however, we might reasonably consider these studies of local American politics in the context of this larger question and reflect on the implications. Development theory of the 1980s – development from below (e.g. Coffey and Polese, 1984; Pottier, 1985; Stohr and Taylor, 1981) – wedded with planning of the 1980s – public-private partnerships and community organization (e.g. Bergman, 1986; Brooks *et al.*, 1984; Davis, 1986) – point to the possibility (although not necessarily the probability) of achieving structural changes at the local level through local processes. Of course, the stimulus and socio-economic prerequisites of endogenous change are far from automatic or even clearly understood. Furthermore, a synthesis of socio-economic and political processes enabling local development is wanting.

While planning for development in the 1980s has largely emphasized socio-economic preconditions and processes, local political processes may also provide insight, if not a vital key to necessary conditions of change. For example, Roberts *et al.* imply in Chapter 10 that family farmers might be able to institute desired changes if only they would address the appropriate issue (structure of production, as opposed to the effects of the existing structure), maintaining a unified front oriented toward long-term as opposed to short-term goals. Although Roberts *et al.* focus on an agrarian context, certainly their study has general implications. In this regard, plans for local development could benefit from adapting the idea of 'appropriate technology' in developing country contexts (Bhalla, 1979, 1984) to 'appropriate organization' in local, grass-roots situations in any development context. Applying this general point to the urban context, Whalley's case study adds an interesting dimension and helps clarify contingent conditions for the achievement of qualitative change: the way in which individuals organize politically around a social or economic issue must be consistent with local political structure and the geography of representation. In addition, the distinction between consumer and producer behaviour addressed by Hodge and Staeheli in Chapter 11, considered in different contexts (agrarian as well as urban), leads us to conclude that structural change requires the spatial separation of these two spheres, in the case that local consumer values conflict with production requirements.

References

Berentsen, W. H. (1981) 'Conflicts between national and regional planning objectives: Austria and East Germany', *Papers of the Regional Science Association*, 48: 135–48.

Bergman, E. M. ed. (1986) *Local Economies in Transition: Policy Realities and Development Potentials*. Durham, N.C.: Duke University Press.

Bhalla, A. S. ed. (1979) *Towards Global Action for Appropriate Technology*, Oxford: Pergamon.

—— (1984) 'Third World's technological dilemma', *Labour and Society* 9: 323–32.

Brooks, H., Liebman, L. and Schelling, C. eds. (1984) *Public-Private Partnership: New Opportunities for Meeting Social Needs*. Cambridge, Mass.: Ballinger.

Burnham, W. D. (1987) 'The turnout problem', in A. J. Reichley, (ed.) *Elections American Style*, Washington, D.C.: The Brookings Institution, 97–133.

Coffey, W. J. and Polese, M. (1984) 'The concept of local development: a stages model of endogenous regional growth. *Papers of the Regional Science Association* 55: 1–12.

Davis, P. ed. (1986) *Public-Private Partnership: Improving Urban Life*, Academy of Political Science, *Proceedings* 36,2.

Erie, S. P. (1985) 'Rainbow's end: from the old to the new urban politics'. in L. Maldonado, and J. Moore (eds.) *Urban Ethnicity in the United States: New Immigrants and Old Minorities*, Urban Affairs Annual Review, 29: Beverly Hills, Cal.: Sage.

Ettlinger, N. (1985) *Towards a Clarification of Regional Economic Change: the United States as a Case Study, 1962–1980*. Ann Arbor, Mich.: University Microfilms.

Hill, D. B. and Kent, M. M. (1988) 'Election demographics', *Population Trends and Public Policy* 14: Washington, D.C.:Population Reference Bureau Inc.

Hirschman, A. O. (1981) *Essays in Trespassing*. New York: Cambridge University Press.

Katznelson, I. (1981) *City Trenches: Urban Politics and the Patterning of Class in the United States*. New York: Pantheon.

Pottier, C. (1985) 'The adaptation of regional industrial structures to technical changes'. *Papers of the Regional Science Association* 58: 59–72.

Stohr, W. (1982) 'Structural characteristics of peripheral areas and the relevance of the stock-in-trade variables of regional science', *Papers of the*

223

Regional Science Association. 49: 71–84.

—— and Taylor, D. R. F. eds. (1981) *Development from Above or Below? The Dialectics of Regional Planning in Developing Countries.* New York: Wiley.

Welch, S. and Bledsoe, T. (1988) *Urban Reform and its Consequences: a Study in Representation.* Chicago: University of Chicago Press.

Part IV

Future directions

Chapter fifteen

Electoral geography and the ideology of place: the making of regions in Belgian electoral politics

Alexander B. Murphy

Elections in most liberal democracies are surrounded by intense and prolonged discussion of spatial patterns of support for candidates, parties, and issues. Ideas about electoral patterns in various functional and perceptual regions provide the basis for much speculation about the chances for a candidate's or a party's success. Indeed, maps showing regional support for political figures and issues are part of the electoral ritual regularly celebrated by the media and the participants. Even the studies of academic geographers are occasionally trotted out in connection with the formulation of campaign strategies, the making of projections, and the analysis of outcomes.

The geographical component of electoral discourse reflects and shapes ideas about which places share sufficiently similar characteristics to be grouped together, and the propagation of spatial electoral generalizations can profoundly affect the election process. Many of the geographical conceptions incorporated into electoral discourse concern the actual or likely vote in various formal administrative units, such as counties, communes, cantons, states, or provinces, but generalizations are also frequently made about regions that lack clear formal or functional significance. In the United States, for example, presidential elections invariably incite an outpouring of comment about the strength of the Republicans in the west or the problems a liberal Democrat faces in the south. In the process, perceptual regions rooted in historical experience or cultural understanding are continuously impressed on the minds of active participants in the electoral process.

Unfortunately, the studies of electoral geographers have all but ignored the geographical component of electoral discourse and the underlying perceptions of place it reflects and generates. This lack of concern presumably results from a predisposition towards the study of electoral 'facts' rather than electoral ideologies. Yet it is intuitively obvious that the ways in which spatial electoral patterns are commonly conceptualized and communicated can be of profound importance to the political process. Geographical generalizations influence candidate decisions about where to campaign, party resolutions about the content of platforms, and ultimately voter choices.

The role of place in electoral geography

The relative neglect by electoral geographers of the nature and significance of geographical discourse in election campaigns is but another example of the constraining framework that has characterized much of electoral geography and has tended to distance the work of electoral geographers from that of other political geographers. As Taylor and Johnston (1979) point out, the great preponderance of studies in electoral geography concentrate on identifying voting patterns or establishing geographical influences on voting and representation. The conceptual isolation of work in this tradition arises out of the general failure to connect the results of research to larger political and social processes (Taylor, 1985: 152). Electoral geography studies all too often leave us with some attractive maps and interesting correlations but fall short of any significant analysis of the causal connections among spatially associated phenomena.

Challenges to the narrow and conceptually isolated character of the sub-field of electoral geography have mounted in recent years, with an accompanying growth in emphasis on the role of place or local context in shaping electoral patterns and processes (for example, Johnston, 1986b; Agnew, 1987). Geographers adopting the perspective that experience is fundamentally determined at the local level have provided important insights into the electoral process, but even they have left the conceptual parameters of place largely unexplored. Although place is generally understood to define experience at a scale below the national level (Johnston, 1986a: 76), attempts to delineate the concept more precisely are lacking. Often it is implicitly treated either as a straightforward formal unit or as a complete spatial abstraction. These approaches convey the message that place is either a preordained, unproblematic spatial unit or something that lacks significance in terms of dimension or scale.

In reality electioneering practices and voting behaviour are critically influenced by ideologies of place. Ideas about the spatial extent of place have obvious impacts on the kinds of issues a voter sees as being personally relevant. Moreover, parties and candidates constantly play on assumptions about regional divisions as they seek to sell themselves to the electorate. Voters' and candidates' understandings of place are anything but simple reflections of formal spatial units. As the theoretical literature on regionalism has demonstrated, place is defined by social relations and constantly changes over time and space (Massey, 1978; Markusen, 1987).

Place is an ideological category that, like any other, involves generalizing across other categorical lines. When a candidate makes a pitch aimed at a particular place or a reporter writes about the vote in a particular region, generalizations are being made that do not apply to many of the inhabitants of that place or region. Yet the role of place in political strategy and discourse is a reflection of an evolved understanding of what place is in concrete spatial terms, and it is a force in sustaining conceptions of places with particular dimensions. Place, of course, means different things to different people at different times. But at any given time and in any given location, ideologies of place have important concrete spatial dimensions that define regions of socio-political identification as well as the nature of generalizations across space.

Treating place in terms of both its contingent nature and its concrete impact on electoral behaviour demands a concern with the process by which place becomes socially and politically defined at particular moments and in particular locations, and with the implications of that process. Pursuing this realm of enquiry requires the student of electoral politics to take the geographical element of electoral discourse seriously, for it is through language that ideas about spatial compartmentalizations and their relationships to social and political patterns are developed and conveyed. This is not to suggest that language is independent of social context. Rather my premise is that language arises out of social context, and in turn helps to shape that context (see, for example, Reiss, 1982).

Incorporating the ideology of place into electoral geographic studies presents formidable challenges. The task is greatly complicated by the fact that people have a variety of associations with place at very different scales. Moreover, the realm of ideology is much more elusive than that of concrete voting behaviour. I suspect that electoral geographers have sidestepped the ideology of place less because of its complexity, however, than because ideology is perceived primarily as being a mask obscuring reality. Yet ideology, as it is fostered and perpetuated through language, is itself a force that fundamentally creates and structures social process (Manning and Robinson, 1985: 1–19; Thompson, 1984: 5–6). An examination of the role of geographical perceptions and discourse in Belgian electoral politics reveals the relevance of this point for electoral geography.

Place and politics in Belgium

The broad outlines of Belgium's political structure are well known.

Belgian political life is ruled by three fundamental oppositions: Catholic and anticlerical, labour and employers, and Dutch-speaking and Francophone (Lorwin, 1966: 147). Three parties have dominated Belgian elections during the twentieth century: Catholic (called Christian after 1945), Socialist (called the Workers' Party before 1945), and Liberal. These parties came into being before the linguistic cleavage became a prominent political issue; hence their early development is much more closely associated with religious and class issues than with language (Hill, 1974: 30). During the twentieth century, language issues came increasingly to the fore and became strongly associated with perceptual, and later formal, regional divisions (Murphy, 1988b: 91–124). During the past thirty years a number of new regional language parties have undermined the long-standing tripartite division of Belgian party politics, and the traditional parties have split along ethnoregional lines.

It is taken for granted in many analyses of political developments in Belgium that a fundamental structural element of the electoral system is the greater strength of the Catholic party in Flanders and of the Socialist party in Wallonia. There is such an 'of course' quality to this assumption that a prominent commentator on the Belgian political scene apparently saw no conflict in declaring in the same paragraph: (1) that Belgium's political regions are defined by the fundamental historical association between Flanders and Catholicism, and between Wallonia and socialism; and (2) that language has played only a minimal role in establishing the political regions of Belgium (Hill, 1974: 44). What is forgotten along the way, of course, is that Flanders and Wallonia owe their very existence as regions to the language divisions. Moreover, as data from the last general election before the strong emergence of the regional parties (the 1961 election) reveal, patterns of support for Catholics and Socialists display more of a core–periphery or urban–rural dichotomy than a Flanders–Wallonia split (Fig. 15.1).

I am certainly not suggesting that the association of Catholicism with Flanders and Socialism with Wallonia is an unimportant feature of political life. These associations are indeed at the heart of a number of important political and electoral developments in modern Belgium. They help explain the lack of correlation between working class votes and support for the Socialists in certain areas, and they lie at the heart of the division of the traditional parties along language lines. But as Fig. 15.1 reveals, there is a very imperfect correlation between language region and party support. Hence, if we expect to gain much insight into the spatial dynamics of the electoral system, we must examine the processes by which regional political associations came about. In other words, we must not always start our analysis of electoral patterns from unexamined assumptions about political regions.

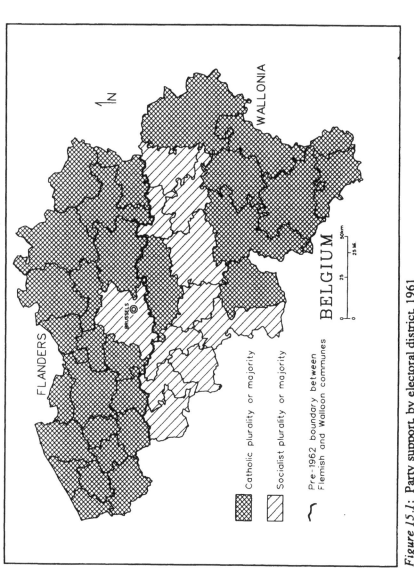

FLANDERS

WALLONIA

N

BRUSSELS

BELGIUM

0 25 50km

0 25 Mi.

▨ Catholic plurality or majority

▨ Socialist plurality or majority

⌒ Pre-1962 boundary between
   Flemish and Walloon communes

Figure 15.1: Party support, by electoral district, 1961
Source: Annuaire Statistique de la Belgique (1962)

The basis of an association between politics and language region

The constitution of the newly independent Belgian state in 1831 allowed for significant autonomy at the local (communal) level (Lorwin, 1966: 150). Flanders and Wallonia had no significance as regional units at the time, and few thought of language distributions as the basis of distinct regional units until the beginning of the twentieth century (Murphy, 1988a: 140–42; De Schryver, 1981: 22–31). Moreover, there could be no easy association between region and language because even though Dutch (or Flemish dialects) predominated in the north and French in the south, a small but powerful political and social elite in the cities of northern Belgium spoke French. In fact, under the suffrage laws of the newly independent Belgian state, voting power was vested in the Francophone population, even in the north.

It is beyond the scope of the present study to examine in detail the rise of the Flemish movement in Belgium (see Elias, 1963–5). Briefly, the movement developed in response to the disadvantaged position of Dutch or Flemish speakers in Belgian society and, during the nineteenth century, focused primarily on the right of individual Flemings to use their language in the courts, the army, and in government. Many of the priests of northern Belgium came to play a significant role in the Flemish movement, in part in reaction to the idea that French was becoming a vehicle for the spread of immoral ideas (Lorwin, 1966: 160; Claeys, 1980: 174). As a result, it was primarily through the Catholic party that the Flemish question was brought before political decision-making authorities, and the association between the Catholic party and the Flemish movement did not go unnoticed (Clough, 1968: 131). In the absence of any strong territorial component to the Flemish movement at the time, however, the association remained weak. This is revealed in the complete absence of references to the political characteristics of the language regions in the campaign propaganda or election reports of the late nineteenth century.

The Socialist party did not come into being until 1885, and its initial successes came in the large cities of northern and southern Belgium, as well as in the Sambre–Meuse industrial district, which lies primarily in the Francophone part of the country. Flemish Socialists focused mainly on socio-economic and suffrage issues in the nineteenth century, which meant that the Socialist movement was not strongly associated with the language cause *per se* (Clough, 1968: 133–4). Since the greatest concentration of industry was located in part of southern Belgium, however, it is not surprising that Socialist gains were greatest among Francophones. Nevertheless, this did not take on regional significance until well into the twentieth century.

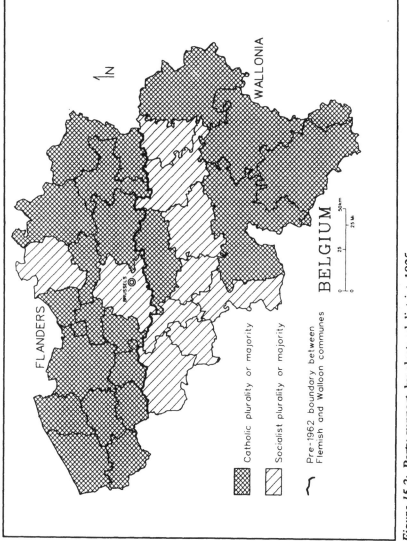

Figure 15.2: **Party support, by electoral district, 1925**
Source: De Smet *et al.* (1958)

The early decades of the twentieth century saw a shift in the Flemish movement from a focus on individual language rights to an emphasis on regional linguistic rights (Murphy, 1988b: 91–123). In response, a nascent Walloon movement also took up the regional language issue. It was during this period that Flanders and Wallonia began to acquire widespread perceptual significance as territorial units, and even assumed a degree of formal administrative significance with the adoption of language laws in the 1920s and 1930s (Maroy, 1969). New geographical generalizations entered political discourse along with this transformation, most notably the equation of Catholicism with Flanders and Socialism with Wallonia.

What was remarkable about the new geographical characterizations is that they were at best only crudely supported by underlying electoral patterns. As data from the 1925 general election reveal (Fig. 15.2), patterns of support for the major parties bore no closer relationship to the language regions than they did in 1961. Certainly there was a Catholic plurality in the north and a socialist plurality in the south, but a number of other social or economic bases might have been chosen for spatial generalizations about party strength that would have yielded much stronger correlations. A core–periphery division would have been the most obvious of these. Hence, to explain the new geographical element in political discourse, we must look beyond simple electoral patterns to emerging regional ideologies.

The traditional political parties were well established before the linguistic issue became a major regional issue, and their bases of support long transcended linguistic lines. Beginning in the early twentieth century, however, the growing regional character of the language issue introduced a new level of place that attracted people's loyalties. In response to the failure of the individual rights approach to the Flemish problem and the need to acquire a broader base of support for their cause, leaders of the Flemish movement began to push the idea of a unified and autonomous Flanders, thereby promoting a sense of both Flemish and Walloon ethno-regionalism. In this atmosphere, the previously described loose ties between the Flemish movement and Catholicism, and between Francophones and Socialism, provided a potential framework for generalizations about the political orientations of the emerging language regions.

These generalizations were encouraged by the greater success of the Socialist party in the industrialized parts of southern Belgium than in their northern counterparts. The early decades of the twentieth century were a period of rapid growth for socialism in Belgium. The greater success of the movement in the south can be attributed in part to the fact that Catholicism was much stronger in the industrialized parts of northern Belgium than in the southern industrial districts (Lorwin, 1966:

156). It was not simply religious doctrine, however, that served as a barrier to the Socialist message in certain Flemish areas. The early 1900s were a time of rapidly growing Flemish ethnic consciousness, and the failure of the Socialist movement to incorporate more Flemish workers within its ranks was tied to the subordination of class conflict in parts of northern Belgium to the goal of Flemish emancipation as a people (Renard, 1961: 228–9). The greater success of the Socialist movement in the south during a delicate period in the development of ethno-linguistic consciousness certainly helped to strengthen the idea that the Socialist party was a Walloon party.

It must be stressed that spatial generalizations about political orientation took on a north–south character only because of developing ideas about regionalism at the time. The economic regions were far from simple reproductions of the language regions (Claeys, 1980: 184), and Catholicism remained strong in significant parts of southern Belgium, as did Socialism in parts of northern Belgium (Fig. 15.2). But with the transformation of language issues into territorial issues, the identification of political orientation with even a segment of a language group could be (and was) interpreted in regional terms. The dissemination of ethno-regional generalizations about political orientation was greatly encouraged by the rhetoric of Socialist and Catholic party leaders during the early decades of the twentieth century.

On the Socialist side, Emile Vandervelde and Jules Destree, prominent party leaders from Wallonia, were vocal in their opposition to many of the demands of the Flemish movement and, in Clough's (1968: 134) words, 'remained champions of French culture and of the French language'. In response, many of the Socialist party's leaders from Flanders took the position that they were concerned with economic issues as opposed to language issues. Given the greater visibility and power of the Walloon Socialists, many Flemings came to believe that the party would be more likely to side with Francophone Belgians on issues of a strongly ethno-linguistic character. This attitude served to affirm the notion of the Socialists as a Walloon party.

The articulated orientation of prominent leaders of the Catholic party was quite different. Under the leadership of Frans van Cauwelaert after World War I, elements within the Catholic party became very strongly associated with the demands of the more moderate wing of the Flemish movement (Clough, 1968: 223). Van Cauwelaert's control of two major Belgian newspapers, *De Standaard* and *De Morgenpost*, helped disseminate the perception of a strong relationship between the Catholic party and the Flemish question and, by extension, Flanders itself.

During the first third of the twentieth century, a host of social and quasi-political organizations came into being, ranging from unions to political action groups. Most of these organizations were identified with the so-called political families (Catholic, Socialist, and Liberal) represented by the major political parties. The growing connections that people were drawing between the Catholic party and Flanders, and between the Socialist party and Wallonia, even if not entirely borne out by actual voting patterns, undoubtedly played a significant role in the stronger development of Socialist organizations in Wallonia and of Catholic organizations in Flanders. The rise of these organizations further abetted the emerging ethno-regional basis of political generalizations.

The organization of workers into different unions in the 1920s and 1930s was of particular significance. The Catholic unions were more successful in attracting members in Flanders, whereas Socialist unions predominated in Wallonia. In part this was a consequence of the developing regional–political associations described above, but it also played a fundamental role in perpetuating those associations. From this time onward, the working classes of Flanders and Wallonia were exposed to different political messages at the local level, and this was to produce different responses to the economic problems that Belgium faced in subsequent decades. This process was further encouraged by other organizations specifically designed to promote a socio-political– ethno-regional bond, including the Katholieke Vlaamsche Landsbond (Catholic Flemish League) and the Katholieke Christen Volkspartij van Vlaanderen (Catholic Christian People's Party of Flanders).

Any ambiguity in the association of the Catholic party with Flanders during the first three decades of the twentieth century was erased in the minds of many during the battle to enact major language laws in the early 1930s. Members of the Catholic party were primarily responsible for applying pressure for the adoption of the laws (Curtis, 1971: 483) and, as the language laws were framed in territorial terms, the relationship between the Catholic party and Flanders was further cemented. Support by Flemish Catholics for the laws led to growing dissension among Walloon Catholics during the 1930s, prompting a short-lived split of the party into linguistic wings in 1936 (Lorwin, 1966: 162).

A further indication of the territorial direction that the language question had taken was the decision to begin compiling official statistics by language region in 1932 (McRae, 1986: 45). The new data provided a statistical basis for generalizations about language regions. Reports of electoral patterns were no longer based primarily on communal, district, or provincial units but on the language regions themselves. Hence it became increasingly easy to gloss over the strong support for the Catholic party in certain Walloon districts (such as the province of Luxembourg or the southern districts of the province of Namur), or for the Socialists in Flanders (such as in the corridor between Brussels and Antwerp). In essence, place was being defined by the social and political developments of the time, and the regional generalizations about electoral patterns that followed were rapidly becoming embedded in the national psyche.

The long-term impact of the politics/language region association

German occupation in World War II brought considerable disruption to the political system, but political developments after the war bore the distinct imprint of the regional–political associations that had developed earlier in the century. In the first decade after the war, politics were dominated first by the dispute over whether the king should step down in the light of allegations of collaboration with the Germans during the war (the so-called royal question), and then by an intense dispute over state funding for private Catholic schools (the so-called school wars). Although neither of these issues directly concerned the language division, the debate over the royal question was framed largely in ethno-regional terms. In fact, it was the failure of the king to receive a majority of votes in Wallonia that eventually led to his abdication (DuRoy, 1968). The battle over state funding to Catholic schools was less divisive from a regional standpoint, but even this issue did not totally escape regional dichotomization, with support for the funding of Catholic schools being more strongly identified with Flanders than with Wallonia (Dunn, 1970: 169). When specifically ethno-linguistic concerns once again surfaced as the premier political issue around 1960, the electorate was predisposed to think of political and electoral patterns in ethno-regional terms.

The similarity in voting patterns between 1925 (Fig. 15.2) and 1961 (Fig. 15.1) is almost certainly tied to the success of the particular geographical discourse documented above. Between 1925 and 1961 significant changes in the socio-economic structure of Belgium had occurred that might have been expected to account for some discernible shifts in the patterns of party support. Most notably, with the movement of the country's economic centre of gravity to the north, the percentage of Flemings employed in the secondary sector increased from 46.0 per cent to 49.9 per cent between the 1920 and the 1961 censuses, whereas the corresponding figures for Wallonia show a drop from 57.8 per cent to 49.6 per cent (Hill, 1974: 32). Theories equating class structure with party support would predict a growth in Socialism in Flanders and a decline in Wallonia, yet, as Figs. 15.1 and 15.2 indicate, the only shift in party plurality or majority at the electoral district level was the loss of one formerly Socialist Flemish district to the Catholics. The language region/political party association provides a telling explanation of this seeming anomaly.

The impact of the political party/language region association extended beyond simple voting patterns. The growing salience of the language question in the early 1960s led to a series of laws partitioning Belgium along language lines. The resulting administrative regions, Flanders, Wallonia, and Brussels, became the basis of the restructuring of the political system through constitutional revisions in 1970 and 1980. During this period language and regional issues were frequently at the centre of the political stage, and the traditional parties could not avoid the centrifugal pressures of a membership holding widely divergent positions on many salient issues. As a result, separate linguistic wings formed within the Catholic (now Christian) party by 1965 and within the Socialist party in 1966. These parties formally divided, in 1969 and 1978 respectively (Fitzmaurice, 1983: 144–61).

The division of the traditional parties is often attributed to the role of the newly created regional language parties in pushing ethno-linguistic issues to the forefront of politics. It is certainly true that the regional parties turned up the heat on language issues, thereby promoting a certain degree of polarization in the traditional parties. But it is also clear that the association of the Catholic party with Flanders, and of the Socialist party with Wallonia, played a role. This is revealed in the approaches taken by the Flemish wing of the Socialist party and by the Francophone wing of the Catholic party in the 1968 election campaign, a campaign charged with linguistic issues. A collection of political posters and leaflets from that election (Dewachter, 1970) reveals that the Flemish Socialists and the Francophone Catholics emphasized their different stances on ethno-regional issues from their Walloon Socialist and Flemish Catholic counterparts, presumably to combat the notion that a vote for the Socialists was a vote for Wallonia, and a vote for the Catholics was a vote for Flanders. For example, many of the posters of the Walloon Catholics asserted strong opposition to some of the fundamental demands of the Flemish movement, including the division of the Catholic University of Louvain along language lines and containment of the expansion of French around Brussels (Dewachter, 1970: 5, 96). At the same time, Socialist Flemings made much of their commitment to removing the Francophone section from the Catholic University of Louvain and to preserving Flanders from Francophone expansion around Brussels (Dewachter, 1970: 106).

Circulation figures for newspapers provide a further indication of how strong the Catholic Flanders/Socialist Wallonia groupings had become by this time. Most newspapers in Belgium follow the same ideological breakdown as the traditional parties: Catholic, Socialist, and Liberal. Catholic newspapers have long dominated much of Belgium and have continued to increase their market share. In a study of newspaper circulation between 1958 and 1974, however, Luyks (1975) showed very different trends between Flanders and Wallonia in the

Catholic and Socialist presses. In Flanders, readership of Catholic papers jumped from 57.5 per cent to 67 per cent, whereas in Wallonia readership increased only marginally from 27.1 per cent to 29 per cent. During the same period, readership of Socialist papers in Flanders declined from 13.8 per cent to 8.7 per cent, with virtually all readers switching to the Catholic press. In Wallonia, by contrast, a more modest drop of 4.1 per cent in readership of Socialist newspapers was largely absorbed by neutral papers. Although changes in newspaper circulation are as much a cause as an effect of language region/political party associations, the Flemings and the Walloons are certainly being confronted with very different messages.

With the division of the traditional political parties along language lines, the electoral system was fundamentally recast. Although the wings of the traditional parties continue to co-operate on a variety of socio-economic issues, there has also been a growing tendency in recent years for members of the now autonomous branches of these parties to go their own separate ways (Murphy, 1988b: 167). Consequently, the long-standing association of Flanders with Catholicism and of Wallonia with Socialism may play a less dramatic role in the future. There can be little question, however, that the development and promulgation of these associations have fundamentally affected the dynamics of Belgium's electoral geography throughout much of the twentieth century.

Conclusion

The electoral process in liberal democracies is an important social ritual in which participants and observers frequently focus on the spatial and social characteristics of place. How place is defined, and why it becomes incorporated into political discourse, is far more than a curious example of geographical thinking. It has significant ramifications for how the electoral process functions. The unfolding of political–territorial developments in Belgium provides a particularly dramatic example of this point. It is only by reference to the long-standing association of Catholicism with Flanders and of Socialism with Wallonia that one can make sense of such seeming anomalies as the low support for the Socialist party among the rapidly expanding working class of Flanders or the relative unimportance of core–periphery distinctions in a country that, on the basis of raw electoral statistics, appears to be an ideal candidate for core–periphery dichotomization.

In cases without such clearly traceable developments in regional thinking, the relationships between political orientation and conceptions of place may be more subtle. Yet the ease of obtaining data based on administrative or electoral districts must not blind us to the fact that place can be, and is, defined at other levels, with enormous implications for the electoral process. If we are to integrate electoral geography successfully into the political geographic enterprise, we must confront the ways in which the electoral process reflects and structures the political organization of space. Taking place seriously, not just in terms of its abstract qualities but also in terms of its ideological dimensions, is an important aspect of that undertaking.

References

Agnew, J. A. (1987) *Place and Politics: The Geographical Mediation of State and Society*. Boston, Mass.: Allen and Unwin.

Annuaire Statistique de la Belgique (1962) Ministère de l'Intérieur, Brussels: Institut National de Statistique.

Claeys, P. H. (1980) 'Political pluralism and linguistic cleavage: the Belgian case', in S. Ehrlich and G. Wootton (eds.) *Three Faces of Pluralism: Political, Ethnic and Religious*. Westmead: Gower, 169–89.

Clough, S. B. (1968) *A History of the Flemish Movement in Belgium: a Study in Nationalism*. New York: Octagon Books.

Curtis, A. E. (1971) 'New Perspectives on the History of the Language Problem in Belgium', unpublished Ph.D. dissertation, University of Oregon.

De Schryver, R. (1981) 'The Belgian revolution and the emergence of Belgium's biculturalism', in A. Lijphart (ed.) *Conflict and Coexistence in Belgium: the Dynamics of a Culturally Divided Society*. Berkeley, Cal.: University of California Press, 13–33.

De Smet, R. E., Evalenko, R. and Fraeys, W. (1958) *Atlas des Elections Belges, 1919–1954*. Brussels: Université Libre de Bruxelles, Institut de Sociologie Solvay.

Dewachter, W. (1970) *Recueil de documents relatifs à la propagande des partis politiques aux élections legislatives du 31 Mars 1968*. Brussels: Institut Belge de Science Politique/Belgisch Instituut voor Wetenschap der Politiek.

Dunn, J. A., Jr. (1970) 'Social Cleavage, Party Systems and Political Integration: a Comparison of the Belgian Swiss Experiences', unpublished Ph.D. dissertation, University of Pennsylvania.

DuRoy, A. (1968) *La Guerre des Belges*. Paris: Seuil.

Elias, H. J. (1963–5) *Geschiedenis van de Vlaamse Gedachte, 1780–1914*. 4 vols. Antwerp: Nederlandsche Boekhandel.

Fitzmaurice, J. (1983) *The Politics of Belgium: Crisis and Compromise in a Plural Society*. London: Hurst.

Hill, K. (1974) 'Belgium: political change in a segmented society', in R. Rose (ed.) *Electoral Behavior: A Comparative Handbook*. London: Collier Macmillan, 29–107.

Johnston, R.J. (1986a) 'Placing politics', *Political Geography Quarterly* Supplement to 5: S63–S78.

—— (1986b) 'Places, campaigns, and votes', *Political Geography Quarterly*

Supplement to 5: S105–S117.

Lorwin, V. R. (1966) 'Belgium: religion, class and language in national politics', in R.A. Dahl (ed.) *Political Opposition in Western Democracies.* New Haven, Conn.: Yale University Press, 147–84.

Luyks, T. (1975) 'De Openierichtingen in de belgische dagsbladpers'. *Res Publica* 17: 223–44.

—— and Platel, M. (1985) *Politieke Geschiedenis van Belgie van 1944 tot 1985.* Antwerp: Kluwer.

McRae, K. D. (1986) *Conflict and Compromise in Multilingual Societies: Belgium.* Waterloo, Ont.: Wilfrid Laurier University Press.

Manning, D. J. and Robinson, T. J. (1985) *The Place of Ideology in Political Life.* London: Croom Helm.

Markusen, A. (1987) *Regions: The Economics and Politics of Territory.* Totowa, N.J.: Rowman & Littlefield.

Maroy, P. (1969) 'L'Evolution de la législation linguistique belge', *Revue du Droit et de la Science Politique en France* 82: 449–501.

Massey, D. (1978) 'Regionalism: some current issues', *Capital and Class* 6: 106–25.

Murphy, A. B. (1988a) 'Evolving regionalism in linguistically divided Belgium', in R.J. Johnston, D. Knight, and E. Kofman (eds.) *Nationalism, Self-determination, and Political Geography.* London: Croom Helm, 135–50.

—— (1988b) *The Regional Dynamics of Language Differentiation in Belgium: A Study in Cultural Political Geography.* Chicago: University of Chicago Research Series in Geography, No. 227.

Reiss, T. J. (1982) *The Discourse of Modernism.* Ithaca, N.Y.: Cornell University Press.

Renard, A. (1961) 'A propos d'une synthèse applicable à deux peuples et à trois communautés', *Synthèses* 186: 204–36.

Taylor, P. J. (1985) *Political Geography: World-Economy, Nation-State, and Locality.* London: Longman.

—— and Johnston, R.J. (1979) *Geography of Elections.* New York: Holmes & Meier.

Thompson, J. B. (1984) *Studies in the Theory of Ideology.* Oxford: Polity Press.

Chapter sixteen

Regulating union representation elections: towards a third type of electoral geography

Gordon L. Clark

Free, uncoerced elections are counted by many commentators as a crucial indicator of the vitality of any democracy. Without regard to the results of elections, a fair electoral process is thought to be an essential ingredient for a just society. Indeed, the fairness of the electoral process is often invoked as evidence in debate over the legitimacy of the state; its actions, and its continued existence as a social institution. Voter fraud, the coercion of voters, and duplicity by candidates in election campaigns (just three of many nefarious possibilities) are taken to be serious violations of the ideals (and ideology) of democracy, even if election outcomes were to be consistent with voters' expectations. In instances where violations of the norms of election campaign conduct are so widespread that they threaten the very integrity of the electoral process, counting outcomes (votes) may be simply a cynical exercise in power and domination.

Given these observations about the conduct of electoral democracy, it is surprising that there are so few studies in geography of the integrity of electoral processes in any country. Electoral geographers seem content to count votes, describe the patterns, and debate the relative contribution of place, social structure, and local milieu in explaining observed patterns. While election outcomes are no doubt very important (and something that I have been concerned with in the context of labour relations), the integrity of the electoral process itself is arguably as important. In this sense, electoral geographers ought to be just as concerned with the regulation of election conduct as they are with the spatial patterns of election outcomes.

Having said this, it is clear that there are few studies in the literature which might serve as guides to expanding the domain of electoral geography. One goal of this chapter is to map-out a shift of emphasis in electoral geography from simply analysing the patterns of outcomes to combining an interest in those patterns with a concern for the regulation of electoral conduct. Specifically, I argue for a greater appreciation of the role of government agencies in regulating elections. The empirical focus of the chapter is the United States' union representation election process, rather than partisan political elections. I begin with a brief overview of the union election process and then move to a discussion of the literature of electoral geography. Based upon recent work on union elections, I draw out implications from this material for the study of electoral regulation.

Union representation elections

Practically all my work in electoral geography has been on the union representation election process. In Clark and Dear (1984) we did acknowledge, though, that electoral democracy is often invoked as a language or legitimizing device for state power. We noted also that electoral democracy is more often a normative ideal than a practical reality. And yet for all its elusiveness in reality it remains as a vital component of Western societies. In this context, the starting point of my work on union representation elections was an observation made by Representative Mead (D.–N.Y.) in 1935 that set out one of the essential normative goals of the Wagner Act: industrial democracy.[1] Mead said, in part, that the Act 'creates a democracy within industry which gives to our industrial workers the same general idea of freedom which the founding fathers conferred upon citizens of the United States'.[2]

Whatever the pattern of outcomes of the union representation process, the ideal of industrial democracy remains a vital ingredient of US labour relations. Thus regulation of the conduct of union representation election campaigns is a topic of considerable interest to workers, unions, management, and the federal government (to cite just four of many possible interested parties). Even though regulation of union elections is often contested by different groups, the regulatory process including the National Labor Relations Board (NLRB; the federal government agency responsible for regulating the election process and adjudicating disputes that arise from election campaigns) is thought to be a distinctive and positive attribute of the US labour relations system compared to European labour relations systems (Bok, 1971).

Representation elections are the normal way local units of functionally similar workers gain or lose union representation; these elections are also the normal way unions add new members to (or subtract existing members from) their national organizations. For individual unions and companies, regulations governing the electoral process and the policing of rules and procedures in these elections are important considerations in their day-to-day conduct. To the extent that government regulation of the conduct of union representation elections affects the outcome of elections, such regulation directly affects the relative strength of organized labour in the US economy. In an era when the American union movement is in decline (relatively and absolutely), understanding the patterns and determinates of union elections with respect to the possible role of government regulation in this decline is a vital and contentious topic of research (Clark and Johnston, 1987a).

There can be no doubt about the importance of a geographical perspective in understanding the union representation election process. In one sense it is a very 'local' phenomenon; being based on specific plants and units of plants it involves people and personalities, their social relationships, and their roles as workers and managers. Local union representation campaigns are the building blocks of union power in areas, industries, and corporations. Failure at this level to organize labour impoverishes (materially and figuratively) the union movement. At the same time, there are often national organizations implicated in any local representation campaign including the organizing agents of unions, anti-union management relations firms whose business is to fight local representation campaigns on behalf of employers, and the NLRB.

To illustrate the kinds of issues adjudicated by the NLRB and the extent of its involvement in local disputes, just consider for the moment a recent case – *Fermont, a Division of Dynamics Corporation of America* (1987).[3] In this case the Board affirmed an administrative law judge's decision overturning the outcome of a representation election arguing that the company's actions had so tainted the electoral process that a fair election was impossible. The company had fired workers for supporting the union's electioneering activities, it had singled out for discipline pro-union workers for supposed poor work performance, it had interrogated workers about their views on union representation, and it had run a competition for an anti-union slogan with prizes including a colour TV set and a microwave oven. In this last instance, the company held a mass meeting of all workers on the morning of the election to announce entries and winners. The NLRB held that the competition was an obvious instance of 'buying endorsements' and as such illegal.[4]

Three types of electoral geography

This case is just one of many (Weiler, 1983). To understand its significance in the context of NLRB case law, and the practicalities of the union representation election process in general, would require a great deal more analysis than can be accommodated in this chapter (see Clark, 1989b). Despite the fact that few academics outside US labour relations specialists are familiar with the union representation election process, it provides a useful reference point for understanding the standard fare of research in electoral geography.

In the electoral geography literature there appear to be two dominant types of research. One type we might term *performance-oriented*; it includes those studies concerned with documenting the performance of classes, groups, coalitions (and many other significant social and spatial 'associations') of voters in elections at various geographical scales. At issue in this type of research is the relative ability of groups of voters to translate their political interests into parliamentary power. While typically focused upon established groups and parties, performance-oriented electoral geography could also analyse the emergence and institutional formation of inchoate political movements and then their entry into the normal channels of electoral representation (a study of the parliamentary mobilization of the 'Green' environmental movement of western Europe would reasonably fit this mould).

This type of research is more empirical than theoretical. Mapping and accounting for the differential geographical performance of contestants in the electoral process is a vital part of electoral geography. Estimating the relative contribution of place-specific variables and attributes compared to socio-economic status variables that are assumed to have standard national interpretations is one expression of empirical agenda of performance-oriented electoral geography. Another expression is attempts at discerning the stability or otherwise of cleavages and schisms that are social and geographical in origin. For example, researchers have been interested in the stability of the north–south split in the popularity of the Democratic party. More recently, geographers have documented the development of a north–south split in Great Britain in the electoral performance of the Labour and Conservative Parties (Johnston *et al.*, 1988).

With respect to union representation elections, many studies have been undertaken of the determinants of union victory and defeat. There are far fewer studies of the electoral geography of union elections, though this is changing. We showed in Clark and Johnston (1987a, 1987b, 1987c, 1987d, 1987e) that there were historically distinctive spatial patterns of union support in America, as well as distinctive spatial patterns in recent union victories and losses. Our study was based upon data collected by the NLRB on the electoral performance of two unions (the International Brotherhood of Electrical Workers and the United Auto Workers) over the period 1970 to 1982 (these data were partially updated in Clark, 1989a). Determinants of union victory and loss were shown to be a combination of local variables, standard wage and employment variables, and national factors including the structure of the electoral process. These papers demonstrated the 'power' of a geographical perspective for understanding the patterns of union election outcomes, a perspective that was lost over the years through theoretical abstraction and ignorance.

The second major type of research in electoral geography is *design-oriented*. Here the research issue is the optimal shape or character of the electoral system. Who is eligible to vote, when do they vote, where do they vote, and how are voters organized into geographical precincts or representative units are essential questions that indicate the breadth of this type of research. Not surprisingly, in electoral geography a great deal of attention has centred upon the proper definition of voting unit boundaries. The definition of boundaries, their inclusion or exclusion of certain well-defined groups (blacks and whites, poor and rich, Democrats and Republicans for example), as well as their implied weighting of individual votes, are all topics that fit well within design-oriented research in electoral geography.

This type of research is more theoretical than empirical. It tends to begin from quite abstract or ideal notions of the theory of democracy, the proper foundations for informed voter choice, and the optimal design of electoral systems. By assumption, this kind of research tends to treat democracy as a universal ideal, even if its canons are violated in practice. Good examples of this kind of research are the studies of voting rights, boundary definition, and electoral apportionment which depend upon statute and court suits for their ideal reference points.

Again with respect to union representation elections, there has been little work in electoral geography on the design of the union electoral system. In fact, there has been little work on this issue in any other discipline. Weiler's (1983) argument against the current US union electoral system (citing its failure to protect workers from dismissal for union-organizing activities), and his argument in favour of the Canadian system, is one exception. He suggests that the electoral system itself is an important contributing factor to the decline of organized labour in the United States. This theme is evident in my own work on the decline of organized labour (Clark, 1989a). I argue that the decentralized character of the union electoral system allows for management subversion of the norms of election conduct as well as division of workers' interests on the basis of community loyalties rather than workers' collective relations with management. Having recognized the need for a basic revision of the design of the electoral system does not, however, guarantee there is a political constituency for redesign.

There is a third type of research in electoral geography that has been heretofore practically ignored; it could be termed *regulation-oriented*. Whereas the two dominant types of research in electoral geography focus respectively on electoral performance and the design of electoral systems, the third type focuses upon the conduct of elections. There are a variety of reasons for neglect of this third type of research in electoral

246

geography. Advocates of performance-oriented research tend to imply that only outcomes matter; either the process is so obvious that it does not require study, or the significance of outcomes in the political arena is such that few people care about the niceties of election conduct. Advocates of design-oriented research tend to assume that once designed, the electoral process is unproblematic. Neither assumption is warranted in the union representation process; outcomes are regularly overturned for violations of the norms of election conduct (Clark and Johnston, 1987d). And there is a great deal of debate over the design of the union representation election process because of the problematic nature of its ideal image.

Regulation-oriented research in electoral geography includes (but is not limited to) the regulation of campaigns, regulation of the conduct of elections, and regulation of electioneering activities. Of course, it must be acknowledged that the three types of research in electoral geography identified in this chapter are all related. It is difficult to make fundamental divisions between types, especially when it is quite evident that any analysis of electoral systems requires an overlapping analytical and empirical strategy which combines all types of research. Nevertheless, in the literature on electoral geography there are few studies that could be termed regulation-oriented. As we have observed, most are performance-oriented. In the following sections I sketch elements of a regulation-oriented approach to electoral geography emphasizing again my recent work on US union elections.

Regulating election outcomes

There are two ways of dealing with regulation. One is to question the goals of regulation; the other is to analyse the practice of regulation. In this section we shall consider the former issue which is often thought to be a political one (that is, conceived in the public legislative arena), whereas the latter issue is thought to be an institutional issue (that is, the actions of the administrative apparatus). In combination, these two elements of regulation are argued by Dworkin (1986) to refer to the integrity of law. I would similarly suggest that these two elements also refer to the integrity of rule-based regulation (Clark, 1985).

Why do governments regulate the conduct of elections? The standard answer is immediately obvious – regulating elections ensures the integrity of democracy. In the context of union representation elections, one need only invoke Representative Mead's comment of 1935 that the object of the Wagner Act was to give industrial workers those same democratic rights held by citizens, to understand how easily debate over the goals of regulating elections is foreclosed by reference to the democratic ideal. While it is obvious that workplace democracy is a desirable goal, it is not at all obvious that the goals of regulation are so neatly summarized by the democratic ideal. In fact other goals are served by the Wagner Act; even elections serve a variety of goals. Ambiguity about the purposes of regulating elections is part of an intense political drama (Clark, 1988).

There are about three different kinds of goals commonly assumed to be served by American labour law and the union electoral process in particular. For economists like Dunlop (1987) and Marshall (1987) American labour law and institutions like the NLRB were designed to foster economic growth. They believe that the collective bargaining system embedded in the Wagner Act is an essential medium for labour–management co-operation and that labour–management co-operation is a necessary ingredient for economic growth. For those who doubt these claims, Dunlop and Marshall refer to the industrial relations turmoil that preceded the passage of the Act in 1935 as well as more recent events including strikes and major labour–management disputes. By this interpretation of the Wagner Act, industrial democracy as conceived by the union representation election process is an *instrument* that promotes collective bargaining, and is thus desirable for its beneficial (albeit indirect) effects on economic growth.

A related economic interpretation of American labour law implies there are limits to industrial democracy, even limits to the desirability of union representation. The Reagan NLRB in a number of important decisions regarding the extent of management discretion in plant closures and the relocation of work argued that the integrity of the collective bargaining system was a secondary consideration when compared to the property rights of management (Clark, 1986, 1988).[5] In this interpretation of the Wagner Act, union representation, collective bargaining, and labour–management co-operation are of *incidental* importance compared to maintenance of the rights and powers of management. By this interpretation, the NLRB's role is to ensure economic efficiency in resource allocation. Once this is assured, the NLRB might reasonably pursue incidental goals, like industrial democracy, assuming that these incidental goals do not impair the overarching goal of economic efficiency.

Of course, there are those who believe that an *essential* goal of the Act was to foster industrial democracy and that the union representation election process was the basic mechanism for achieving that goal. In a previous section I quoted a Congressman of the time who said just that; that industrial democracy was a worthy goal in its own right. Indeed, by Representative Mead's interpretation, fostering industrial democracy was the overriding goal of the Wagner Act. Notice, though, that this is just one of a number of interpretations of the Act and of the significance of union representation elections in the Act. Without going into more details of interpretations, theories, and policies, it is apparent that union representation elections might be interpreted as an *essential* part of the Act, an *instrument* of the Act, or just an *incidental* part of the Act.

During the Reagan administration the NLRB went from being dominated by Carter appointees (1980–1) to being dominated by Reagan appointees. In this time the Board's interpretation of the significance of union representation elections went from assuming these elections to be an essential part of the Act to being just one of a number of incidental aspects. The choice of emphasis was a political one. The Board inherited a range of interpretations of the significance of union elections. Given its mandate (as part of the Reagan administration) to improve economic efficiency and the background and experience of most appointees in management-oriented labour law firms, it is not surprising that it peripheralized the imperatives of industrial democracy. Of course, there was a great deal of political dispute over its choice of emphasis, as there was dispute over the administrative performance of the NLRB in general. This is what I mean by claiming that the significance of union representation elections is part of an intense political drama.

So far we have treated the NLRB as the sole legitimate agency responsible for regulating and administering federal labour law. Yet there is a great deal of dispute between federal courts, the Board, and the US Department of Labor over their respective powers in the area of labour law and labour policy (Clark, 1989c). The federal courts have a powerful position relative to the Board because the courts are an option for challenging Board decisions. As well, the Board depends upon the courts for enforcement of its orders.

Just as the Board must choose among competing interpretations of the significance of industrial democracy, so too must the courts. The federal courts do not always agree with the Board's interpretations (even though there is a presumption that the Board has special expertise in this area). The Department of Labor has become more involved in facilitating labour–management co-operation, circumventing (but not altogether avoiding) the Reagan NLRB. Rather than use representation elections as a device for mobilizing labour, the Department of Labor has become involved in joint labour–management teams that are less about democratic representation and the election process, and more about fostering economic efficiency through worker involvement and responsibility for increasing productivity. In these ways, representation elections may have even less significance for federal labour policy than implied by the Board.

My argument in this section had two parts. First, it should be apparent that the regulatory process need not have a stable or unambiguous goal. Even though industrial democracy is rhetorically appealing, bolstered by political debate and argument at the time of the passage of the Act, this does not automatically mean that it is in practice the essential goal of regulation. The goals of regulating union representation elections vary with the changing balance of political forces. Second, the significance of union representation elections for federal labour policy may vary as different institutions pursue their own versions of labour policy. There may be other ways of achieving the goals of labour policy that do not require union elections.

Rules and procedures of elections

Having considered the goals of regulating elections in the context of union representation, I now turn to the practice of regulation. A naive empiricist might claim that studying the rules and procedures that govern the practice of elections is unimportant compared with election outcomes because the rules and procedures are relatively stable and well known. This is implausible on a number of grounds. With respect to union representation, all the evidence suggests that the rules and procedures governing the conduct of elections are not so stable as one might imagine (Clark, 1989b; Clark and Kim, 1988). Moreover, following Wittgenstein (1953) there are good theoretical reasons to expect that rules and procedures are not stable over many cases because they require the texture of events, different circumstances, to determine the meaning and application of these rules. In this sense, rules do not carry a universal meaning that transcend context.[6]

To illustrate this issue, we might pause for a moment and consider the question of the proper definition of rules and procedures. Should rules and procedures be national in scope or should they recognize regional differences in the customary practice of labour relations? One answer would be that since federal labour law pre-empts state and local laws, the rules and procedures of representation elections should be national not local. While this answer suits federal interests in regulating labour relations, it need not be thought the final answer. Historically, the south has resisted the implementation of federal labour law; indeed the southern Congressional delegation (Republican and Democrat) has been an unwilling partner in recent attempts at labour law reform (Clark, 1989a). If rules and procedures are dependent upon the context of union representation elections, there is a real question about the degree of respect accorded to different contexts.

This is a question about standards; their definition and application. It is also a question about geography and the relevance of a set of local standards in relation to other local standards and national standards. It appears that in regulating union representation campaigns and elections, the NLRB has applied national standards derived from northern union environments to southern elections. In this way the NLRB has attempted to nationalize local labour–management relations (with arguably little success).

If this appears too abstract, or too obscure, it is because the rules and procedures – the formal mechanisms by which union representation elections are regulated – are themselves problematic and open to multiple interpretation. Take, for example, the NLRB's rules on campaign propaganda (see Clark, 1989b). These rules have gone through a variety of incarnations since 1945. Over most of the 1980s the Reagan NLRB dominated the definition of campaign propaganda, arguing that there is no such thing as propaganda, though there is a possibility of deliberate misrepresentation by witholding information. Generally, a majority of the Reagan Board believed propaganda to be of little consequence when voters (workers) make their choices in elections (between union representation and non-representation). Nevertheless, the use of language in representation campaigns remains a vital and contested topic, involving a great deal of litigation from the Board through to the federal courts.

In some court cases the question of propaganda is raised when the company involved threatens to relocate production and close the facility if workers vote for union representation. This threat was apparently voiced by management in the Fermont case. According to many unions, this kind of threat is illegal, or should be illegal. They believe that it unfairly puts workers in the position of choosing between union representation and their continued employment. The Board, though, has to judge the significance of this threat before deciding to overrule an election outcome and order a new election. What standards does the Board use? What would be the appropriate standards? For some members of the Board, this kind of threat is transparently empty. They would apply a universal standard of adjudication – that threats of relocation have little force because employees can always obtain other jobs (assuming the spatial labour market system is efficient and transaction costs including search costs are trivial).

Less sanguine members of the Board argue that these kinds of threats ought to be evaluated according to the actions and past behaviour of the company. By these terms, propaganda is judged according to local circumstances. What may be an idle threat in some places and some firms need not be an idle threat in others. Thus the meaning of a threat, the significance of a threat, and the definition of appropriate remedial action all depend upon context. And yet it is also apparent that the NLRB has not always treated localities equally. The local context of any threat or regulation may not be equally respected; the Board may apply the standards of one context to other contexts depending upon its experience in different settings.

The rules and procedures of union representation elections are complicated and prone to litigation by unions and management. If ideally unimportant compared to election outcomes, practically these regulations are at the very heart of dispute between electoral contestants. Basically, the rules and procedures can have significant effects on the conduct of election campaigns, and may have just as important ramifications on how electoral contestants value election outcomes.

One way of illustrating this point is to return again to the debate over the causes of the declining electoral performance of American unions (documented in Clark, 1989a). If we simply assume the rules and procedures of union elections to be unproblematic in application and adjudication then unions' declining electoral performance might reasonably be interpreted as the result of changing voter (worker) preferences. By this logic, workers do not value unions as they once did. This interpretation is sometimes offered by management organizations in lobbying efforts against labour law reform. They use data on unions' declining electoral performance to 'show' the irrelevance of unions for the average worker. Even the union movement has at times accepted this interpretation as a rationale for a new form of unionism, one which sells unionization as a way of improving consumption standards as opposed to shop-floor representation.

Alternatively, if we accept that the rules and procedures of election campaigns are problematic, declining electoral performance could be interpreted as a problem of regulation, not a problem of declining popularity. If regulation of elections is so problematic, it may be more reasonable to claim that the true preferences of workers for union representation are being frustrated by the inadequacies of regulation. By this interpretation, management subversion of the election process and subversion of the NLRB appeal process may be aided and abetted by a Board unwilling or unable to recognize the substantial impact it may have on union electoral performance through its regulations. This interpretation is offered by Weiler (1983) and others as a critique of current forms of electoral regulation and as a critique of the Reagan administration. One consequence of this critique has been a steady stream of recommendations for labour law reform – specifically reform of the union representation election process.

Avenues of future research

In emphasizing the importance of analysing the regulation of elections, I do not mean to imply that either of the other two types of electoral geography is in any sense less important than previously imagined. Rather, my goal has been to map out a broader domain for electoral geography, a domain that includes the institutions that regulate elections as well as the patterns of election outcomes. As I attempted to show, understanding the electoral performance of election contestants as well as understanding the design of the electoral process itself cannot be separated from the regulatory process. In these ways electoral geography can be both broader and of more consequence for public policy.

I would suggest that an appreciation of government regulation of elections is vital for a number of reasons. Most significantly, an appreciation of the regulatory process is a step away from the idealization of democracy. Any reading of the related literature in geography and political science would seem to suggest that democracy is in fact as it is in principle. That is, there is a sense in which emphasis on outcomes implies that democracy as a process functions as it was designed in theory. This is not a credible assumption, as most students of electoral geography would readily admit. Nevertheless, to ignore the practice of democracy as evident by government regulation of the election process (one dimension among many) tends to reinforce the idealization of democracy as if it functions as designed. Thus an emphasis on regulation is a step towards understanding the practice of democracy.

My approach to regulation is premised upon a set of embedded methodological assumptions. It is important that we recognize these assumptions explicitly because they may not be shared by all readers or all researchers in electoral geography. A crucial assumption is that the outcomes of elections are not self-evident in their meaning.[7] To understand their significance requires an appreciation of how they were achieved, the role of regulation, and the goals of the regulatory process. Surely, as someone might protest, election outcomes are self-evident in that they are the essential material of democracy. Perhaps. But elections can be both more and less than democracy – as I tried to indicate in my discussion of the multiple goals embedded in the union representation election process. Not all goals need respect or even significantly value election outcomes even though the regulatory process was nominally designed to ensure its integrity.

A second assumption is that the value of an electoral outcome or set of outcomes is not independent of how it was achieved. Of course, it is obvious that outcomes and procedures are intimately related. Any design of an electoral system must begin from a sense of what are desirable and possible electoral outcomes. I accept this idea. However, my point goes further than the design process and assumes there is an inherent normative value we assign to electoral outcomes on the basis of how they were achieved in practice. In this sense, we value outcomes to the extent that the electoral process was fair and genuinely acceptable to the electoral contestants. Regulation has a vital role to play in ensuring the integrity of the electoral process. Another, more formal, way of expressing this same idea is to say that ranking outcomes from least to most desirable is insufficient to determine the 'best' outcome; we may value more a less desirable outcome than a more desirable outcome because the regulatory process was fair and equitable (see Scheffler, 1982 on the theory of consequentialism).

A third assumption is that formalism – regulating the conduct of elections by rules and procedures – is problematic. As a practical reality, rules and procedures are often ambiguous and variable in meaning and in their application by regulatory agencies. To imagine otherwise is to idealize regulation and the regulators. 'Superman' was a comic-book hero; there is no reason to imagine that comic heroes are located at crucial locations in the regulatory process. Practical reality is one explanation of the problematic status of rules and procedures. A more theoretical argument is that because rules and procedures do not exist as universals any more than outcomes have a meaning in themselves, they depend upon events and circumstances for substantiation. Thus it is inevitable that rules and procedures are variable in meaning. In this sense, regulation, like interpretation, is always a contest of meaning(s) and interests (Norris, 1985).

These assumptions may appear unrelated to the aforegoing discussion of union representation elections. In case this is true, it is worth restating some of the conclusions of the previous sections of this chapter. An important conclusion was that the regulatory process is believed by many to be implicated in the declining electoral performance of American unions. Weiler (1983) and Clark (1989a) for different reasons argue that the NLRB's conduct of union representation elections is fundamentally flawed. Implied in this conclusion are the three assumptions noted above. First, understanding the significance of electoral outcomes requires an appreciation of the NLRB's goals and objectives in the regulatory process. In recent years it appears that the NLRB has not valued the electoral process as highly as other aspects of its regulatory mandate. Second, the value we attach to the outcomes of the union representation process has steadily declined over the past decade as evidence has been introduced to show that the regulatory process has been subverted by management. Third, it has become obvious that the rules and procedures regulating the conduct of representation elections are unstable and variable in application. This is obvious in litigation over the application of the NLRB's rules regarding campaign propaganda.

Electoral geography has been dominated by outcomes and design considerations. A broader and more mature electoral geography should integrate these considerations with an appreciation of the regulation of elections.

Notes

1. Wagner Act of 1935, known officially as the National Labor Relations Act, as amended, 29 U.S.C. 141–144, 151–187 (1982).
2. Congressional Record, vol. 79, p. 9710, 19 June 1935.
3. Fermont, a Division of Dynamics Corporation of America and Teamsters 1040, a/w International Brotherhood of Teamsters, AFL–C10, 286 NLRB No. 96, 30 December 1987.
4. Specifically, the NLRB held that 'it is obvious that the respondent was buying endorsements, with a promise of material awards of significant monetary value. Such a grant of benefits to selected individuals, prior to the election, has a tendency to interfere with the employees' Section 7 rights' (pp. 3–4).
5. See Milwaukee Spring Division of Illinois Spring Co. II, 268 NLRB 601 (1984), and Otis Elevator, a wholly owned subsidiary of United Technologies II, 269 NLRB 891 (1984).
6. See Clark (1985) and Dworkin (1978) for a critique of the stability of rules in law and judicial adjudication.
7. Putnam (1987) makes this same point in a more abstract manner. Referring to realism, he argues that to speak about a 'thing in itself' without reference to its relative position in a system or language of signification is to talk about an empty concept.

References

Bok, D. (1971) 'Reflections on the distinctive character of American labor laws', *Harvard Law Review* 84: 1394–463.

Clark, G. L. (1985) *Judges and the Cities: Interpreting Local Autonomy.* Chicago: The University of Chicago Press.

—— (1986) 'Restructuring the US economy: the National Labour Relations Board, the Saturn project, and economic justice', *Economic Geography* 62: 289–396.

—— (1988) 'A question of integrity: the National Labor Relations Board, collective bargaining and the relocation of work', *Political Geography Quarterly* 7: 209–28

—— (1989a) *Unions and Communities under Siege: American Communities and the Crisis of Organized Labor.* Cambridge: Cambridge University Press (forthcoming)

—— (1989b) 'The local context of federal regulation: the language of US union representation campaigns', *Transactions*, Institute of British Geographers.

—— (1989c) 'Dispute over the powers and relative autonomy of the National Labour Relations Board', *Policy Studies Journal* (forthcoming).

—— and Dear, M. (1984) *State Apparatus: Structures and Language of Legitimacy.* Winchester, Mass., and London: Allen and Unwin.

—— and Johnston, K. (1987a) 'The geography of US union representation elections 1: The crisis of US unions and a critical review of the literature', *Environment and Planning* A 19: 33–57.

—— and Johnston, K. (1987b) 'The geography of US union representation elections 2: Performance of the United Auto Workers union and the International Brotherhood of Electrical Workers union, 1970–82', *Environment and Planning A* 19: 153–72.

—— and Johnston, K. (1987c) 'The geography of US union representation elections 3: The context and structure of union electoral performance (the International Brotherhood of Electrical Workers union and the United Auto Workers union, 1970–82)', *Environment and Planning A* 19: 289–311.

—— and Johnston, K. (1987d) 'The geography of US union representation elections 4: Patterns of close elections and determinates of the margins of victory and loss (the International Brotherhood of Electrical Workers union and the United Auto Workers union, 1970–82)' *Environment and Planning A* 19: 447–69.

—— and Johnston K. (1987e) 'The geography of US union representation elections 5: Reconceptualizing the theory of industrial unionism'. *Environment and Planning A* 19: 719–34.

—— and Kim, J-H. (1988) 'Challenged ballots in US union representation elections'. *Working Paper* 88–9. Pittsburgh, Pa. School of Urban and Public Affairs, Carnegie Mellon University.

Dunlop, J. (1987) 'The legal framework of industrial relations and the economic future of the United States', in C. Morris (ed.) *American Labor Policy*. Washington DC: Bureau of National Affairs, 1–15.

Dworkin, R. (1978) *Taking Rights Seriously*. Cambridge Mass.: Harvard University Press.

—— (1986) *Law's Empire*. Cambridge, Mass.: Harvard University Press.

Johnston, R. J., Pattie, C. J. and Allsopp, J. G. (1988) *A Nation Dividing? The Electoral Map of Great Britain 1979–1987*. London: Longman.

Marshall, F. R. (1987) 'The Act's impact on employment, society, and the national economy', in C. Morris (ed.) *American Labor Policy*. Washington, D.C.: Bureau of National Affairs, 16–34.

Norris, C. (1985) *Contest of Faculties: Philosophy and Theory after Deconstruction*. London: Methuen.

Putnam, H. (1987) *The Many Faces of Realism*. LaSalle, Ill.: Open Court.

Scheffler, S. (1982) *The Rejection of Consequentialism: A Philosophical Investigation of the Considerations underlying Rival Moral Conceptions*. Oxford: Clarendon Press.

Weiler, P. (1983) 'Promises to keep: securing workers' rights to self-organization under the NLRA', *Harvard Law Review* 96: 1769–827.

Wittgenstein, L. (1953) *Philosophical Investigations*. trans. by G. E. M. Anscombe, London: Macmillan.

Chapter seventeen

Extending the world of electoral geography

Peter J. Taylor

In the *Geography of Elections* (Taylor and Johnston, 1979) we attempted to overcome the uncoordinated nature of much electoral geography research by setting our study within the social cleavage theory of Stein Rokkan (1970). This volume continues in that tradition. There is an important limitation to this approach, however, as some reviewers pointed out. The contents of the *Geography of Elections* had a huge geographical bias towards the 'First World'. This volume is equally susceptible to such criticism. Our parochialism is not due to a dearth of elections outside our First World since there have been literally hundreds of such elections over recent decades. But electoral geographers have chosen, by and large, not to study these particular elections. It is the contention of this chapter that this limitation of coverage has been a prime cause of the poverty of theoretical development in our sub-discipline: we need to extend the world of electoral geography.

The broader theoretical framework of world-systems analysis enables us to interpret our geographical bias and understand what we have been studying as part of a wider political world. Electoral geographers have chosen to be students of liberal democracy. Elections failing to conform to this particular ideal have been largely ignored. Hence liberal democracy has been treated as a 'given', an unproblematic form of state. The unwritten assumption is that liberal democracy is the 'natural' culmination of political progress. That it produces a much preferable politics to what occurs outside the First World today is not the issue here; the point is that current electoral geography provides a very partial understanding of elections even in liberal democracies. Studying parties and voting patterns in the USA but not in Mexico is not a very sensible way of producing a body of knowledge about elections.

Our argument will proceed in three stages. In the first we describe the nature of liberal democracies and the context of their success. They are very recent phenomena of the world political scene and we shall argue that they have become a viable form of state only when combined with social democracy and its politics of redistribution. This provides an explanation for the geography of liberal democracies with its concentration in the rich core zone of the world-economy. The First World bias of electoral geography is a side-effect of this pattern.

In the second section we rediscover the problematic nature of liberal democracy. Our uncritical acceptance of this state form has meant that fundamental political dilemmas in its development have been forgotten or ignored. This has resulted in a failure properly to understand the nature of today's elections beyond the liberal democracies of the core. This is, therefore, another reason why electoral geography has such a spatial bias – it cannot cope adequately with elections outside its current affluent world.

It follows that our call to extend the world of electoral geography is not a call for additional case studies of 'Third World' elections. Without adequate theory to handle these further studies the result would be a return to the uncoordinated character of previous electoral geographies. In the final section, therefore, we use world-systems analysis to provide a theoretical framework that can cope with elections in both liberal democracies and other political contexts. We will argue that by concentrating on case studies at the state level most electoral geographies have dealt with contingent social relations at the expense of the necessary social relations that underwrite every election. Our conclusion is that there is no way forward in electoral geography until we recognize both contingent and necessary relations in our studies. This is why we need to extend the world of electoral geography.

The triumph of liberal democracy

It is the thesis of this chapter that liberal democracy in combination with social democracy has provided the ideal state form for resisting the communist threat to the capitalist world-economy. The outbreak of the Cold War in Europe made this state form politically necessary, and the post-World War II Kondratiev growth phase made it economically possible. Hence the new 'affluent society' of the West included two political visions – liberal freedoms and social justice. We shall term the result the 'liberal-social-democratic state'. It is this state that has been the subject matter of electoral geography.

Of course this new state form was not created by Americans and West Europeans out of thin air. They were able to harness existing social and political processes in a new combination for the new Cold War circumstances. The concepts of social democracy and liberal democracy both have their origins in political processes in the core of the world economy in the nineteenth century. They relate to three separate questions confronting nineteenth-century politicians. First, there was

258

the constitutional question which primarily concerned the liberals. The replacement of relatively arbitrary power by constitutional checks and balances was their answer. Second, there was the political question, which was posed by the democrats: who was to hold the power in the constitutional states if not the people as a whole? Third, there was the social question, exploited by the socialists; what can be done about the new poverty of the people of the cities? The first two questions and answers were originally opposed to one another – the liberals were not wrestling for control of the state to give it away to other political groups. Hence the concept of liberal democracy is not a product of the nineteenth century. On the other hand social democracy was a common idea in the nineteenth century. The democratic and social questions were typically connected in radical programmes of social and political change. Most Marxist parties before 1917 were 'Social Democratic Parties'. But the meaning of this social democracy has changed since 1917, and certainly since 1945. It is now the most common label for non-Marxist socialist parties. If we concentrate on the period since 1945 we see that liberal democracy and social democracy have come together as alternative descriptions of the politics of the states of the core of the world-economy. These transformations are part of reason for the ambiguity that so bedevils the concept of democracy. Let us consider each concept in its post-1945 phase.

Liberal democracy

Liberal democracy has come into its own since 1945. In the Cold War it has become the political ideal that is offered to combat the totalitarian foe. Hence it is much more than a label for a party or a policy; it describes a type of state.

Liberal democratic states are usually characterized by three basic properties: (1) *universal suffrage* meaning that all adult citizens are entitled to vote in elections; (2) *pluralistic elections* meaning that there is competition between two or more parties to form the government which is settled by elections; and (3) *political freedoms* meaning that citizens are free to associate and form any party, to campaign and speak on behalf of any party and of course to vote for any party in elections (Johnston, 1989).

These properties are found with only minor blemishes in all core countries at the present time. One or more of the properties are also often found in non-core countries. This has led some researchers to devise measures of the 'degree' of liberal democracy for all countries. The problem with this sort of cross-sectional exercise is that the properties can easily be varied by governments (elections can be declared or cancelled) and often are in non-core countries. If we treat the concept as describing a type of state, however, we are implicitly assuming some degree of stability of the existence of liberal democratic properties. We may suggest that we can be sure that we are experiencing a liberal democratic state only when its properties have survived through at least two generations of politicians. Core states since 1945 meet this additional criterion. Almost all non-core states do not. Some have experienced 'liberal democratic phases' in a far more unstable state form. We should not confuse such 'phases' with a state 'type' as cross-sectional studies are apt to do (e.g. Coulter, 1975; Bollen, 1983).

Social democracy

As in the case of liberal democracy, we will interpret social democracy to mean more than a party or policy. In the Cold War competition it became the 'social progressive' alternative to communism in the core. I will use social democracy to mean a type of state commonly referred to as a welfare state.

Social democratic or welfare states are usually characterized by three major properties: (1) *welfare rights*, meaning that all citizens have automatic rights to a wide range of social services and support provided by the state; (2) *social democratic consensus*, meaning that all major competitors for government accept the need for historically large welfare expenditure; and (3) *redistribution*, meaning that the welfare is paid for largely through progressive taxation.

These properties are far less precise than those defining liberal democracy but this reflects the more complex nature of social democracy. The particular social services and supports provided vary greatly among the core states; compare, for instance, Sweden and the USA. Nevertheless we can identify the creation of a new general type of state formed in the middle part of this century where intervention in social problems became legitimated and normal. Whether it was called 'New Deal' or 'democratic socialism' we can identify it as the culmination of social democratic pressures from the late nineteenth century onwards. In the Cold War era all major parties in the core are social democratic even in the 1980s (e.g. Margaret Thatcher's 'The National Health Service is safe in my hands') and elections only concern the degree of social services and supports not whether there should be such state intervention.

The liberal-social-democratic state

Today all liberal democracies are social democracies. We are looking at the same state from two different angles. Each 'democracy' sustains the other. The particular outcome, the liberal-social-democratic (LSD) state, represents an all-round triumph.

Viewing this state from the perspective of nineteenth-century politics, the descendants of the liberals now have their 'freedoms' constitutionally guaranteed: for the democrats their one person/one vote ideal has become accepted as the proper way of conducting elections, whereas the socialists have their social question firmly at the top of the political agenda. No wonder the LSD state has proved the best bulwark against the attractions of communism for working people in western Europe. A liberal democratic consensus built upon a social democratic consensus has squeezed communist parties out of contention for power in the core. This was a necessary requirement for having the Cold War.

The LSD state as an anti-communist instrument did not just conveniently 'emerge' at this time, it had to be created. In the USA this is symbolized by the rout of Henry Wallace, the Progressive Party candidate for the US presidency in 1948. Wallace came fourth. He had been Roosevelt's popular vice-president from 1940 to 1944; in 1948 he was supported by the Communist Party. The anti-communism was consolidated in the following years by McCarthyism. Similar, if less public, political clearances were being made in other Western states in the name of anti-communism (Whitaker, 1984). More generally, the USA was contributing directly to producing a 'safe' politics in western Europe and Japan (Kolko and Kolko, 1973). The end result was a politics appropriate to the Cold War: a domestic settlement between capital and labour which included a bipartisan foreign policy to combat communism. This is what the LSD state provided.

Liberal social democracy as the dominant political creed of the core of the world-economy is now forty years old. The LSD state has proved its stability and resilience over both the growth and the stagnation phases of the fourth Kondratiev cycle. Its success in this one corner of the world has blinded many politicians and political scientists to its innate limitations: the LSD state has become an ideal to be transferred to the Third World to further the crusade against communism. A new phrase was coined, 'the Free World', as an umbrella term to cover non-communist regimes everywhere. The optimism of building such a world beyond the core of the world economy was supported by political scientists who built developmental models where liberal democracy was the automatic product of 'modernization'. But we now know this was not to be. It is the non-communist countries of the Third World that have given the lie to such optimistic political theories. Liberal democracy is not transferable; it is not unproblematic. Indeed there was nothing inevitable about it even in the core: in the nineteenth century it was not an outcome that most politicians thought possible. We must return to that period to remind ourselves of the dilemmas facing politicians a century ago in order to understand the limitations of liberal democracy today.

The hidden dilemmas of liberal democracy

The politics of the modern world system do not run smoothly. The system is not a machine that operates to produce pre-determined outcomes. Rather constraints are imposed and opportunities provided in an uneven pattern over space and time, in which politicians compete with one another between and within states. The result is the variety of dilemmas which is the real world of politics.

Liberals against democracy

Liberalism as a movement treated the constitutional problem of the modern state. As we have seen, its relation to the political problem was much more problematic. Constitutional principles based on the 'people' turned the traditionally feared mob into citizens. As citizens they should presumably have political rights to go with their new political responsibilities. Such logic led most liberals in the nineteenth century to assume that universal suffrage (or at least manhood suffrage) would eventually come about. But that day represented 'the great fear' (Arblaster, 1984). Surely the lower classes would take the opportunity to attack property and privilege?

Interestingly enough the evidence did not support these fears. The early test case was the USA where white manhood suffrage from 1828 onwards produced the Jacksonian democracy that de Tocqueville analysed. If this was Europe's future, all was not doom. But in 1848 the real test came for European liberals as they had to choose between democratic forces and reaction when the revolutions of that year came to a head. From the initial Paris repression onwards, liberals solved their dilemma by choosing reaction over democracy (Arblaster, 1984). No thoughts of liberal democracy here. And this political analysis of the liberals was supported by their great foe Marx. He understood that the class contradictions made any combination of democracy and liberalism an unstable state form. The capitalist state's standard form would have to be authoritarian, as in Louis Napoleon's French solution (Przeworski, 1980).

This analysis was soon found to be incorrect. In Britain the 'leap into darkness' occurred in 1867 with the enfranchising of most of the urban male working class. Civilization did not end. The state continued and found a fundamental new source of legitimacy. Liberal democracy was now a possibility. The experience was repeated in other countries of Europe and North America.

Why were the original analysis and dire predictions so wrong? The question can be answered at two levels. First, in terms of the practical politics of elections the whole process of mobilization was underestimated. People do not automatically vote in their own interest. They have to be 'educated' to their interest and convinced that voting is a way of promoting that interest. It was not a quick fix but a long haul, as Rokkan (1970) has so carefully shown. Mobilizing new voters and changing the political agenda was an enormous task which took generations to achieve.

The second reason for the failure of the dire predictions concerns the relations between the classes described by Gramsci's concept of hegemony. 'The ideas of the ruling class are the ideas of society' to be sure, but it was not a simple matter of false consciousness, a sort of huge hoax perpetrated on a gullible working class. The consent of the working class as reflected in elections and elsewhere has a material base. In capitalist society reproduction of all depends ultimately on reinvestment by capitalists. The entire society is structurally dependent on the dominant class (Przeworski, 1980). Hence it is hardly surprising that the latter's ideas are seen as universal in comparison to the particularist claims of other classes. This is the hegemonic situation within which liberal democracy was to develop. Capitalist behaviour was thus a necessary condition to realize the interests of other classes but it was not sufficient. Elections became the mechanism for introducing the latter's interests. As Przeworski (1980) so cogently puts it 'Democracy is the modern Bonaparte.' And the new interests represented have finally produced social democracy. Only after 1945 was the social question firmly at the top of the agenda so that the LSD state could be finally created. Social democracy proved the liberals' dilemma to be no longer relevant. Liberals are no longer against democracy, at least in the core.

The corollary of 'Free World authoritarianism'

We return again to the non-core zones of the world economy. Here the dilemma remains for liberals although they would claim not to be against democracy but rather against 'populism'. American political scientists have produced their elite theory of democracy to justify just such positions. But generally throughout the periphery and semi-periphery liberals have lost the constitutional battle and so are not a major force in confronting the political and social questions. There can be no true 'free world' of liberal states to confront the communist bloc because the political and social problems beyond the core are insoluble by liberal means. The best that can be achieved, from the core's perspective is for non-core countries to carry on liberal foreign relations, reserving their necessary illiberalism for their domestic population. This is the typical combination favoured by the USA and carried on by its closest non-core allies. The classic cases are the Latin American 'national security states' of the 1960s and 1970s. As in 1848, liberals in the core have been able to overcome their dilemma and accept brutal new regimes in a sort of US-sponsored 'Tyrants for Freedom club'. The fear of the mob is now the fear of communism.

The difference between 1848 and 1988 for liberals is one of geography. In 1848 illiberalism was accepted at home to deter the mob. In 1988 the necessary illiberalism is to be found abroad – there is a neat geographical separation between our liberalism and their repression. This is the fundamental political geography of our age.

It is in the zones beyond the core, therefore, where Marx's 'Bonapartist state solution' commonly occurs (Alavi, 1979). The structural dependence on the dominant class that has produced capitalist hegemonic control in core states has not been repeated elsewhere. In the Third World the most important capitalist interest is usually external to the country. Hence 'The essential problem about the state in post colonial societies stems from the fact that it is not established by an ascendant native bourgeoisie but instead by a foreign imperialist bourgeoisie' (Alavi, 1979: 41). In this situation all interests are special interests. The state exists as a mediator for a fragmented dominant class and the direct producers have no reason to be loyal to the state. It follows therefore that the distinction drawn between authoritarian regimes and elected governments in Third World countries may well be overstated (Heeger, 1974: 71). In First World terms this distinction is crucial; in poor countries, it often only represents alternative means by which local elites organize their control of the state. Elections can mean very different things in different parts of the world.

The party politics of the world-economy

Describing the dilemmas of politicians is a necessary aid to extending the world of electoral geography. But it is not sufficient. We need to make sense of electoral outcomes not as a new set of uncoordinated studies but as part of an overall global process. Thus far we have implied this in our discussion, now we explicitly bring in a world-systems analysis to interpret the world of political parties and elections. We begin by deriving the two basic politics around which parties have been organized and elections fought. These are then interpreted in electoral geography terms by identifying tendencies towards congruent and non-congruent electoral politics.

Deriving two politics

We shall begin by outlining the materialist basis of Wallerstein's (1979, 1983) world-systems analysis. This will be a minimum sketch to provide the concepts for deriving the two politics of the world economy (Wallerstein, 1983, 1984).

The prime motor of change in the modern world-system is ceaseless capital accumulation. This produces uneven development in both space and time generating a core–periphery geography and a Kondratiev-cycle history. This basic process takes place through the operation of two fundamentally opposed classes, the controllers of capital and the direct producers of wealth. Although these classes are system-wide in scope they have organized their politics within and between states. This inter-state system is a prerequisite of capital accumulation. The political 'balkanization' of the world prevents a political direction of the system which would spell doom for the controllers of capital and their ceaseless accumulation: that is to say, it would produce a different prime motive in a different system. Hence the necessary basis of politics in the modern world-system is a combination of inter-class competitions and intra-class conflicts with intra-state debates and inter-state rivalries. Wallerstein (1984) is able to derive two politics from this situation.

In order to accumulate capital all controllers at the end of the day have to buy cheap and sell dear. This can be achieved in one of two ways; either prices are raised or costs are reduced. In the former case prices can be controlled by creating quasi-monopolies. This produces an intra-class politics between controllers of capital and has typically involved using the state to create territorial quasi-monopolies. Hence for every state there is a politics revolving around how that state relates to the rest of the world-economy. Thus international relations will cover policies of war and diplomacy, protection and free trade. There will be competition between controllers of capital within the state over the appropriate policy. This was originally represented in court politics, often organized through secret conspiracies. Later, with the acceptance of legitimate opposition, different interests of capital were represented by loose parliamentary factions. These became parliamentary parties in the mid-nineteenth century when interests became converted into political principles. In Britain, for instance, the Whigs and Tories became Liberals and Conservatives. This party politics continued to revolve around issues of war, imperialism, and trade. The key economic distinction between parties in most countries posited free traders v. protectionists.

Reducing costs to increase capital accumulation involves an entirely different politics. This centres on the inter-class conflict between the controllers of capital and the direct producers. The former have used the state to limit the amount of wealth created that stays with the direct producers. In the nineteenth century political pressure from the direct producers in the form of trade unions challenged the 'natural law' of subsistence wages. Towards the end of the nineteenth century, following suffrage extensions, extra-parliamentary parties were founded to introduce this inter-class politics on to the state agenda.

At the beginning of the twentieth century, therefore, there were two forms of politics in operation (Blondel, 1978). A traditional politics emphasizing international relations was pursued by cadre parties. These parties developed from the old parliamentary parties and involved the addition of electoral committees beyond parliament to campaign and generate support. They are cadre parties because power remained firmly at the centre with the parliamentary leadership; the role of the party in the country is that of a supporters' club. Alongside this old politics a new class politics is emerging, with mobilizing mass parties who are not just campaigning but are promulgating new ideologies to convert people to membership of the party as a movement.

Non-congruent electoral politics before the LSD state

These two politics with their two very different sorts of parties were quite unsuitable material for producing a viable liberal democracy. In the period just before 1914 there was tension between these politics as they existed in uneasy combination (Blondel, 1978).

In terms of electoral geography we can predict what have been termed non-congruent electoral politics (Taylor, 1984). The politics of power remains in the hands of the cadre parties, who pursue their concern for international relations on behalf of their capitalist interests. The extension of the franchise, however, has produced a new mass politics of support among the direct producers that may be unrelated to the politics of power. The classic case is probably in the USA where direct producers were largely voting along ethnic or cultural lines while party elites in Washington D. C. were divided between free trade Democrats and protectionist Republicans (Archer and Taylor, 1981). The geography of support and the geography of power were not congruent.

In Britain this phase of non-congruent politics is represented by Hobson's paradox (Taylor, 1984): in 1910 'Producer England' (the industrial and urban north) voted Liberal and Labour, and therefore for free trade and competition for declining industries, whereas 'Consumer England' (the south) voted Conservative, and therefore for protection and higher prices. This represents an old politics of power (free trade v. protection), still under the control of the old cadre parties, being challenged in the politics of support by a new inter-class politics. This suggests a transition between old and new politics but in fact the latter was not able to triumph until 1945.

The tension between the two kinds of politics continues in the period between the two world wars. In fact the economic problems of the Kondratiev stagnation phase bring the tension to breaking point in many countries. This is a time of retreat for democracy and electoral politics. Mobilizing parties of both right and left prosper at the expense of cadre parties everywhere except the USA. It is at this time that the latter's exceptionalism is confirmed with the restriction of the Progressives to the upper Mid West after earlier defeats of the populist and socialist attempts at mobilization.

One particularly interesting feature of the pre-LSD state period is the lack of contrast between core and periphery in the changing politics. As Wesson (1982: 15) points out, in 1929 all the major Latin American countries had civilian elected governments. They seemed to be following a similar political trajectory to Europe. As economic conditions worsened in the 1930s many civilian governments were deposed but at the time this did *not* indicate a divergence from the European experience. Hence in 1939 approximately half today's core liberal democracies were ruled by dictators. Viewing the world just before World War II, therefore, we see a retreat of electoral politics in the periphery and parts of the core. All the political ideas to produce the liberal social democratic state were available but the material conditions and political circumstances were not yet suitable.

Congruent electoral politics, the LSD state and the politics of failure

It is all change after 1945, as we have seen. In the core class conflict is ameliorated in the new Kondratiev boom conditions. This is directly reflected in changes in the political parties. The cadre parties become more popular in orientation in order to survive and as part of their reconciliation with social democracy. They become representation parties which aim to reflect public opinion. At the same time the mobilizing socialist parties are 'domesticated' to become alternative representation parties. Hence the two types of party have merged into a single form with the cadre parties becoming more sensitive to the voters and the mobilizing parties less ideological. This is the new world of party politics in which pollsters, advertising agencies, and TV images substitute for fundamental policy differences (Taylor, 1986).

The new material conditions that make this new politics possible result in the ditching of the old cadre concern for trade policies. Under US hegemony all core states support free trade, for instance. There is the development of a congruent politics in the sense that the geography of support matches the geography of power. The new class politics of the parties is reflected in the voting mosaics of elections. This even occurs in the USA despite the earlier failure of the socialists – the Democrats became the 'working class party' after the New Deal.

This neat congruent politics of the LSD state in the core has not been repeated elsewhere. After 1945 Latin American politics diverges from the western European trajectory despite common membership of the 'free world'. Cadre parties operating a clientistic politics compete with new mobilizing parties of various populist hues (Mouzelis, 1986). The instability of this combination, found in the core before 1945, continues in Latin America after 1945. Here there is no conversion to a common representation type of party. The material conditions are insufficient to develop a politics of redistribution that can begin to satisfy the mass of direct producers. Instead there is a politics of failure (Osei-Kwame and Taylor, 1984).

A politics of failure means that no elected government will be able to satisfy its supporters: the resources available to the government are just not adequate. The most obvious feature of this form of politics is that governments are regularly voted out of office since the conditions are made for opposition party successes. Dix (1984) and Werz (1987) report that in 44 Latin American elections between 1945 and 1985 incumbent governments have been re-elected on only 11 occasions. In Sri Lanka under its original parliamentary constitution the final six elections led to six changes of government.

A second feature of the politics of failure is that it will be reflected in a non-congruent electoral geography. In Ghana, for instance, parties differing on economic principles have been elected with ethnic geographies of support (Osei-Kwame and Taylor, 1984). But this is not a repeat of the non-congruent politics of the core before the LSD state. Here the geography of support is highly unstable (Johnston *et al.*, 1987). Since parties regularly let their supporters down, they are unable to maintain the continuous long-term patterns of support Rokkan (1970) describes for the core. The result is a complex pattern of ever-changing geographies (for Ghana see Osei-Kwame and Taylor, 1984; for India see Taylor, 1986).

A further feature of the politics of failure is that in cases where parties have been able to maintain power over several elections they have produced another type of party – the aggregative party. This pragmatic and opportunist party does not mobilize support nor represent popular opinion; it is a cadre party that survives by bringing together blocs of votes into temporary coalitions for particular elections. The most important feature of such aggregations is that they should be a majority. The Congress Party in India is the classic case of a successful aggregative party (Taylor, 1986).

A final feature of the politics of failure is, of course, that it produces very unpopular politicians so that the most common outcomes are coups and repression of the popular forces unleashed but unsatisfied by the parties.

This world-systems analysis contradicts the usual Cold War interpretation of world politics that identifies the prime political divide between 'democratic West' and 'communist East'. North–South differences are considered to be largely economic so that the 'South' becomes a stage for East–West political rivalries. In world-systems analysis things are not so clear cut. The liberal social democratic state is to be found only in the 'West' to be sure, but the politics of the communist states of the 'East' are closer to the West than are the authoritarian politics of the South.

Communist states are illiberal but have accommodated the social question into their state form whereas the states of the South are typically illiberal and have been unable to tackle the social question. The similarity between West and East is most clearly expressed in the stability of their regimes compared to the South's excessive instability. These relative differences reflect, of course, the material contrasts between the three 'worlds' with the 'East' ranked between 'West' and 'South' (hence the East is the 'Second World'). The corollary is that if we wish to predict a geographical diffusion of the LSD state it is in the East and not the South where we should be looking for evidence. Enter Gorbechev, farewell Cold War.

A conclusion: from ends to means

In country after country in the Third World as authoritarian governments fail to improve living standards the call goes out for a 'return to democracy'. The problem is, of course, that the particular contingencies that may allow elections to be held in a poor country cannot cancel out the broader necessary social relations that always turn the resulting elected government into a failure. This is often fore-shadowed at election time when death counts accompany vote counts. In many countries, in calling for an election we are asking for a mini-civil war.

Detailed studies of electoral geographies within a world-systems framework can contribute to a body of research that combats the naive assertion of democracy as the 'solution'. This illusion is a product of Cold War thinking where elections were converted by the West from a means to an end to an end in themselves. This was only *after* elections had produced social democracy as an end in the West, of course. As we have seen in the Third World, elections act as a means of legitimating alternative elites in power since social democracy is beyond what they can afford. The question, therefore, is: can elections in the Third World have a purpose that contributes to the emancipation of the peoples of these poorest countries? Defining such an end is the task confronting *all* social and liberal democrats in the world today. More subtly, we need to enquire whether elections in the core are developing new ends beyond social democracy. Extending the world of electoral geography can contribute to understanding the ends to which elections are directed today and their possible alternative purposes in the future.

References

Alavi, H. (1979) 'The state in post-colonial societies', in H. Goulbourne (ed.) *Politics and State in the Third World*. London: Macmillan, 40–59.

Arblaster, R. (1984) *The Rise and Decline of Western Liberalism*. Oxford: Blackwell.

Archer, J. C. and Taylor, P. J. (1981) *Section and Party*. New York: Wiley.

Blondel, J. (1978) *Political Parties*. London: Wildwood House.

Bollen, K. (1983) 'World-system position, dependency and democracy: the cross-national evidence', *American Sociological Review* 48: 468–79.

Coulter, P. (1975) *Social Mobilisation and Liberal Democracy*. Lexington, Mass.: Lexington Books.

Dix, R. H. (1984) 'Incumbency and electoral turnover in Latin America' *Journal of Interamerican Studies* 26: 435–48.

Heeger, C. A. (1974) *The Politics of Underdevelopment*. London: Macmillan.

Johnston, R. J. (1989) 'The individual in the world- economy', in R. J. Johnston and P. J. Taylor (eds.) *A World in Crisis*. second edition. Oxford: Blackwell.

—— O'Neill, A. B. and Taylor, P. J. (1987) 'The geography of party support' in M. J. Holler (ed.) *The Logic of Party Systems*. Dordrecht: Kluwer, 265–7.

Kolko, J. and G. (1973) *The Limits of Power*. New York: Harper & Row.

Mouzelis, N. P. (1986) *Politics in the Semi-periphery*. London: Macmillan.

Osei-Kwame, P. and Taylor, P. J. (1984) 'A politics of failure: the political geography of Ghanaian elections, 1954–1979' *Annals of the Association of American Geographers* 74: 574–89.

Przeworski, A. (1980) 'Material bases of consent: economics and politics in a hegemonic system' in M. Zeitlin (ed.) *Political Power and Social Theory*. Greenwich: Jai Press.

Rokkan, S. (1970) *Citizens, Elections, Parties*. New York: McKay.

Taylor, P. J. (1984) 'Accumulation, legitimation and the electoral geographies within liberal democracy', in P. J. Taylor and J. W. House (eds.) *Political Geography: Recent Advances and Future Directions*. London: Croom Helm, 117–32.

—— (1986) 'An exploration into world- systems analysis of political parties', *Political Geography Quarterly*, Supplement to 5: 5–20.

—— (1987) 'The poverty of international comparisons' *Studies in Comparative International Development* 22: 12–39.

—— and Johnston, R. J. (1979) *Geography of Elections*. London: Penguin.

Wallerstein, I. (1979) *The Capitalist World-Economy*. Cambridge: Cambridge University Press.
—— (1983) *Historical Capitalism*. London: Verso.
—— (1984) *The Politics of the World-Economy*. Cambridge: Cambridge University Press.
Werz, N. (1987) 'Parties and party systems in Latin America' in M. J. Holler (ed.) *The Logic of Party Systems*. Dordrecht: Kluwer, 223–43.
Wesson, R. (1982) *Democracy in Latin America*. New York: Praeger.
Whitaker, R. (1984) 'Fighting the Cold War on the home front', *Socialist Register*, 23–67.

Index

Printed and bound by CPI Group (UK) Ltd, Croydon, CR0 4YY

22/10/2024

01777621-0014